Rapid Assessment Program

A Rapid Marine Biodiversity Assessment of Milne Bay Province, Papua New Guinea— Survey II (2000)

Editors
Gerald R. Allen, Jeff P. Kinch,
Sheila A. McKenna, and Pamela Seeto

Bulletin
of Biological
Assessment

29

Center for Applied Biodiversity
Science (CABS)

Conservation International

Australian Institute of
Marine Science

Western Australian Museum

The *RAP Bulletin of Biological Assessment* is published by:
Conservation International
Center for Applied Biodiversity Science
Department of Conservation Biology
1919 M St. NW, Suite 600
Washington, DC 20036
USA

202-912-1000 telephone
202-912-0773 fax
www.conservation.org
www.biodiversityscience.org

Conservation International is a private, non-profit organization exempt from federal income tax under section 501c(3) of the Internal Revenue Code.

Editors: Edited by Gerald R. Allen, Jeff P. Kinch, Sheila A. McKenna, and Pamela Seeto
Design/production: Kim Meek
Maps: [Map 1, page 169] Mark Denil; all other maps by Albert Kambar

RAP Bulletin of Biological Assessment Series Editors:
Terrestrial and AquaRAP: Leeanne E. Alonso and Jennifer McCullough
Marine RAP: Sheila A. McKenna

ISBN: 1-881173-68-2
Library of Congress Catalog Card Number: 2002111693

The designations of geographical entities in this publication, and the presentation of the material, do not imply the expression of any opinion whatsoever on the part of Conservation International or its supporting organizations concerning the legal status of any country, territory, or area, or of its authorities, or concerning the delimitation of its frontiers or boundaries.

Any opinions expressed in the *RAP Bulletin of Biological Assessment* are those of the writers and do not necessarily reflect those of CI.

RAP Bulletin of Biological Assessment was formerly *RAP Working Papers*. Numbers 1–13 of this series were published under the previous title.

Suggested citation:
Allen, G. R., J. P. Kinch, S. A. McKenna, and P. Seeto. (Eds.). 2003. A Rapid Marine Biodiversity Assessment of Milne Bay Province, Papua New Guinea—Survey II (2000). RAP Bulletin of Biological Assessment 29. Conservation International, Washington, DC, USA.

The Global Conservation Fund, United Nations Development Program, David & Lucile Packard Foundation, and the Henry Foundation generously supported publication of this report.

Using New Leaf Opaque 60# smooth text paper (80% recycled/60% post-consumer waste), and bleached without the use of chlorine or chlorine compounds results in measurable environmental benefits[1] For this report, using 1,404 pounds of post-consumer waste instead of virgin fiber saved…

5	Trees
457	Pounds of solid waste
502	Gallons of water
655	Kilowatt hours of electricity (equal to .8 months of electric power required by the average U.S. home)
830	Pounds of greenhouse gases (equal to 672 miles travelled in the average American car)
4	Pounds of Hazardous Air Pollutants, Volatile Organic Compounds, and Absorbable Organic Compounds combined
1	Cubic yard of landfill space

[1] Environmental benefits are calculated based on research done by the Environmental Defense Fund and the other members of the Paper Task Force who studied the environmental impacts of the paper industry. Contact the EDF for a copy of their report and the latest updates on their data. Trees saved calculation based on trees with a 10" diameter. Actual diameter of trees cut for pulp range from 6" up to very large, old growth trees. Home energy use equivalent provided by Pacific Gas and Electric Co., San Francisco. Landfill space saved based on American Paper Institute, Inc. publication, Paper Recycling and its Role in Solid Waste Management.

Table of Contents

Participants

Gerald R. Allen, Ph. D. (Fishes)
Conservation International
1919 M St., N.W., Suite 600
Washington, DC 20036 USA

Mailing address:
1 Dreyer Road
Roleystone, WA 6111
Australia
Fax: (618) 9397 6985
Email: tropical_reef@bigpond.com

Mark Allen, B. Sc. (Reef fisheries)
1 Dreyer Road
Roleystone, WA 6111
Australia
Fax: (618) 9397 6985
Email: leucopogon@bigpond.com

Douglas Fenner, Ph. D. (Reef corals)
Australian Institute of Marine Sciences
P.M.B. No. 3
Townsville, Queensland 4810
Australia
Email: d.fenner@aims.gov.au

Edward Kibikibi (Community liaison)
Conservation International (PNG)
P.O. Box 106
Waigani, NCD
Papua New Guinea
Email: ci-png@conservation.org

Jeff Kinch, BA Hons 1st Class (Anthropology)
Conservation International (PNG)
P.O. Box 804
Alotau, Milne Bay Province
Papua New Guinea
Email: j.kinch@conservation.org

Tessa McGarry, B. Sc. (Reef ecology)
Conservation Biology Group
Department of Zoology
University of Cambridge
Downing Street
Cambridge CB2 3EJ
United Kingdom

Timothy Werner, M. Sc. (Reef ecology, conservation)
Marine Biodiversity Program Director
Center for Applied Biodiversity Science
Conservation International
1919 M St., N.W., Suite 600
Washington, DC 20036 USA
Email: t.werner@conservation.org

Pamela Seeto, Bsc Honours 1st Class (Marine Ecology)
(Reef ecology)
Regional Advisor–Western Pacific Program
David and Lucile Packard Foundation
P.O. Box 5911, Boroko,
Papua New Guinea
Email: pseeto@packard.org

Roger Steene (Photographer)
P.O. Box 188
Cairns, Queensland 4870
Australia

Emre Turak (Reef corals)
Australian Institute of Marine Sciences
P.M.B. No. 3
Townsville, Queensland 4810
Australia

Fred E. Wells, Ph. D. (Malacology)
Department of Aquatic Zoology
Western Australian Museum
Francis Street
Perth, WA 6000
Australia
Email: wellsf@museum.wa.gov.au

CONSERVATION INTERNATIONAL

Conservation International (CI) is an international, non-profit organization based in Washington, DC. CI acts on the belief that the Earth's natural heritage must be maintained if future generations are to thrive spiritually, culturally, and economically. Our mission is to conserve biological diversity and the ecological processes that support life on earth and to demonstrate that human societies are able to live harmoniously with nature.

Conservation International
1919 M St., N.W., Suite 600
Washington, DC 20036 USA
(202) 912-1000 (telephone)
(202) 912-0772 (fax)
http://www.conservation.org

Conservation International (PNG)
P.O. Box 106
Waigani, NCD
Papua New Guinea
Email: ci-png@conservation.org

AUSTRALIAN INSTITUTE OF MARINE SCIENCE

The mission of the Australian Institute of Marine Science (AIMS) is to generate the knowledge to support the sustainable use and protection of the marine environment through innovative, world-class scientific and technological research. It is an Australian Commonwealth Statutory Authority established by the Australian Institute of Marine Science Act of 1972 in recognition of a national need to manage Australia's marine environment and marine resources.

Australian Institute of Marine Science
Cape Ferguson, Queensland
PMB No 3, Townsville MC QLD 4810
(61-7) 4753-4444 (telephone)
(61-7) 4772-5852 (fax)
http://www.aims.gov.au

WESTERN AUSTRALIAN MUSEUM

The Western Australian Museum was established in 1891, and its initial collections were geological, ethnological and biological specimens. The 1960s and 1970s saw the addition of responsibility to develop and maintain the State's anthropological, archaeological, maritime archaeological, and social and cultural history collections. The collections, currently numbering over two million specimens/arte/facts, are the primary focus of research by the Museum's own staff and others. The aim is to advance knowledge on them and communicate it to the public through a variety of media, but particularly a program of exhibitions and publications.

Western Australian Museum
Francis Street
Perth, WA 6000
Australia
(61-8) 9427-2716 (telephone)
(61-8) 9328-8686
http://www.museum.wa.gov.au

Acknowledgments

The survey and report were funded by the CI Tropical Wilderness Protection Fund (now the Global Conservation Fund) and the United Nations Development Programme. We also acknowledge the generous support of the David and Lucile Packard Foundation and the Henry Foundation.

We are very appreciative of the support and guidance of CI-PNG and particularly Gaikovina Kula, Maureen Ewai, Pamela Seeto, Edward Kibikibi, Jeff Kinch, and David Mitchell. We also thank the former Milne Bay Governor, the Hon. Titus Philemon, and his helpful staff, especially his Provincial Administrator Hauo'fa Sailasa.

We would especially like to express our appreciation to the people and residents of Milne Bay Province for their support and help during the survey. We are grateful to the agencies that helped make this survey possible, especially the Department of Environment and Conservation.

We are indebted to Rob Vanderloos and the crew of the *Chertan* for providing a comfortable and efficient base of operations during the first half of the RAP. Wayne Thompson and the crew of *Marlin 1* provided logistic support during the latter part of the survey.

Executive Summary

INTRODUCTION

This report presents the results of a rapid field assessment of Milne Bay Province, which encompasses the extreme southeastern tip of mainland Papua New Guinea and an extensive offshore area immediately eastward. It covers approximately 265,000 square kilometres, mostly situated in the Solomon Sea, an area heavily dotted with islands and shoals separating PNG from the neighboring Solomon Islands. The province includes three major mainland districts: a 130-km long stretch on the south coast extending between Samarai and Orangerie Bay, Milne Bay proper, and Goodenough Bay, lying immediately northward, and the adjacent southeastern part of Collingwood Bay. Major islands or island groups include the D'Entrecasteaux Islands, Trobriand Islands, Woodlark Island, Egum Atoll, and the Louisiade Archipelago including Rossel Island, Sudest Island, Misima Island, Calvados Chain, Conflict Group, Engineer Group, Sideia Island, and Basilaki Island. In addition, there is a host of widely scattered smaller islands. The entire area is characterized by an extensive and complex system of submerged and emergent coral reefs.

A previous Marine RAP was conducted at Milne Bay Province during October–November 1997. Survey locations included Milne Bay proper, East Cape area, D'Entrecasteaux Islands, Engineer Group, Conflict Group, Nuakata region, and the southern tier of islands including Samarai, Sideia, and Basilaki. Due to the successful outcome of the first survey, the Milne Bay Provincial Government invited CI to undertake an additional survey, with emphasis on sites that were not visited previously. Consequently, the focus of the 2000 RAP was Goodenough and Collingwood bays on the mainland, as well as selected offshore locations including the Amphlett Islands, D'Entrecasteaux Islands, Rossel Island, Sudest Island, and Calvados Chain in the Louisiade Archipelago.

Overview of Marine RAP

The goal of Marine RAP is to rapidly generate and disseminate information on coastal and near-shore shallow-water marine biodiversity for conservation purposes, with a particular focus on recommending priorities for conservation area establishment and management. Marine RAP deploys multi-disciplinary teams of marine scientists and coastal resource experts to determine the biodiversity significance and conservation opportunities of selected areas. Through underwater inventories generally lasting three weeks, Marine RAP surveys produce species lists that serve as indicators of overall biological richness, as well as recording several measurements to assess overall ecosystem health. During each survey, RAP supports parallel assessments of local human community needs and concerns, which become incorporated into the final recommendations.

By comparing the results obtained from many surveys, Marine RAP is ultimately focused on ensuring that a representative sample of marine biodiversity is conserved within protected areas and through other conservation measures.

Milne Bay Province

Milne Bay encompasses the most extensive coral reef area of any province in Papua New Guinea. These reefs are scattered over approximately 265,000 km² of ocean. In spite of its considerable area, Milne Bay Province is one of the least populated provinces, with only 205,000 inhabitants. Although there are several large-scale development projects in the province, including mining and oil palm, most people earn their living primarily by subsistence farming, fishing, harvesting of sedentary marine resources, and the sale of products from these activities. The main challenge in Milne Bay is to improve the services and economic options for its people without sacrificing their cultural identity and unique biodiversity.

The Milne Bay 2000 Survey

The 2000 Marine RAP survey of Milne Bay Province assessed 57 sites over a 26-day period (30 May–24 June, 2001). General site areas were selected prior to the actual survey in order to maximize the diversity of habitats visited, thus facilitating a species list that incorporates maximum biodiversity. Due to logistic problems it was necessary to use a different live-aboard boat for each half of the survey. Because of the smaller size of the second vessel, it was necessary to reduce the size of the RAP team. Consequently, no data were taken for molluscs on this part of the survey. At each site, an underwater inventory was made of two or three faunal groups selected to serve as indicators of overall coral reef biodiversity: scleractinian corals, molluscs (except sites 29–57), and reef fishes. Additional observations were made on the environmental condition of each site, including evaluation of various threat parameters. Observations and data on reef fisheries were also gathered.

The general survey area (see Map 1 on page 169) covered approximately 18,000 square kilometers, encompassing reefs of Goodenough and Collingwood bays on the PNG mainland, as well as Amphlett Islands, D'Entrecasteaux Islands, the Louisiade Archipelago, and isolated sites at Sideia, Basilaki, and Bently islands. Charter dive boats based at Alotau reached the 57 survey sites.

SUMMARY OF RESULTS

Most coral reefs in Milne Bay Province remain in good condition with relatively rich biodiversity. Notable results from the survey include:

- *Corals*: A total of 418 species of scleractinian corals were recorded, which is more than half of the world's species, indicating that Milne Bay Province is truly one of the globe's richest areas for corals. Several potential new species were collected.

- *Molluscs*: Although molluscs were not surveyed during the Louisiades portion of the RAP, a total of 643 species was recorded with a range of 34–119 per site. Com-

bined with the results of the 1996 survey, the current molluscs total for Milne Bay Province is 954 species.

- *Reef Fishes*: A total of 798 species were recorded. The overall reef fish fauna of Milne Bay resulting from the 1997 and 2000 RAP surveys consists of 1109 species, the highest for any area in the Melanesian region. At least one new species (Pomacentridae) was collected.

- *Reef Fisheries:* Significant stocks of edible reef fishes were observed on most reefs, but holothurians and giant clams were often scarce, the result of intensive harvesting. Detailed information for Brooker Island in the Louisiade Archipelago indicates that many local communities are dependent on marine resources for food and income.

- *Reef Condition:* Reefs were generally in good shape with significant amounts of live coral cover. Using CI's Reef Condition Index, it was noted that nearly 50 percent of surveyed reefs were in good, excellent, or extraordinary condition. These are sites with the best combination of coral and fish diversity. They are also relatively free of damage and disease. In contrast, only eight percent of reefs were considered to be in poor condition, but these were mainly confined to sheltered bays with high levels of silting.

CONSERVATION RECOMMENDATIONS

The coral reefs of Milne Bay Province play an integral role in sustaining coastal communities and represent an important component of PNG's rich natural heritage. Until recently, this environment was under minimum stress, mainly due to the small human population of Milne Bay Province and its remoteness. However, there are increasing signs of habitat degradation, mainly due to land-based activities and over-harvesting of sedentary marine resources. For these reasons, and because the province contains some of the best examples of relatively undisturbed reefs in the entire Coral Triangle region, it is vitally important that both government agencies and communities commit necessary resources to ensure that Milne Bay's reefs are conserved for future generations. In order to achieve this aim we propose the following recommendations:

1. **Evaluate and address threats to the marine environment from land-based activities.** The effects of sedimentation on reefs from deforestation, agriculture, and mining are of some concern. Any emerging threats from land-based activities should be closely monitored and appropriate actions taken to mitigate any detrimental effects they may have on the marine environment. Watershed protection also needs to be a primary objective.

2. **Establish community-based marine conservation and resource management areas that result in sustainable fisheries management.** There are numerous tools available to fisheries managers and conservation practioners to achieve this. One of the tools that could be considered by communities and agencies for Milne Bay Province is the establishment of Marine Protected Areas (MPAs). Currently there are no MPAs in Milne Bay Province. The ultimate success of any MPA is dependent upon the recognition of the biological, social, and economic issues relevant to the local communities and their subsequent incorporation into the selection and design process of MPA establishment.

3. **Conduct more scientific surveys to fill gaps in biological and habitat data.** Further biological surveys and baseline data collection are required to prioritize more specific areas for MPA establishment. Milne Bay Province occupies a vast area, and several RAP surveys would be required to adequately cover all the important reef areas. Locations that remain unsurveyed include the Trobriand Islands, Woodlark Island, Egum Atoll, Misima Island, Sudest Island south Barrier reef, Bramble Haven, Long/Kossman reef, and the southern mainland coast.

4. **Continue collaboration with the National Fisheries Authority on rigorous stock assessments of commercially harvested species, and influence the formulation of species management plans.** The over-harvesting of sea cucumbers, giant clam, and shellfish is a serious concern in Milne Bay Province. Rigorous stock assessments are necessary on a continual basis to gauge the current status of these resources and influence the development of appropriate management plans.

5. **Continue to link tourism benefits to the conservation of marine resources.** To achieve the sustainable conservation of marine resources, the benefits from the dive industry should be shared with communities to provide incentives for conservation.

6. **Continue to develop and implement an environmental education and awareness program to impart conservation values to students and communities at all levels.** An environmental education and awareness campaign is required to instill conservation values among students and communities in order to generate support for marine conservation efforts in Milne Bay Province.

7. **Continue participation in the annual PNG Coastal Cleanup campaign.** This campaign was introduced in 1999 as a collaborative effort between CI and the National Capital District Commission (NCDC). This activity increases public awareness of marine issues in PNG, in particular those pertaining to litter and waste (such as plastic bags), and the detrimental effects they are having on the marine environment and marine species.

8. **Assist in the community mapping of resource ownership.** Customary marine tenure gives control and ownership of most near-shore areas, including reefs, to communities. At present customary marine tenure is loosely defined in Milne Bay Province, and this issue needs to be resolved in order to achieve long-term conservation outcomes and avoid conflicts between communities as resources become scarce or as regulations of MPAs are enforced.

9. **Strengthen capacity within the province for effective implementation of the marine conservation program.** Milne Bay Province has very few staff and resources for conservation or the provision of other services. Strengthening the capacity of the Milne Bay Provincial Government is therefore necessary for it to confront the growing environmental pressures and implement an effective marine conservation program.

10. **Enforce existing laws and propose options for surveillance of illegal foreign fishing vessels.** Greater enforcement of regulations outlined by species management plans is required to address the ongoing problem of over-harvesting and to ensure the long-term viability of these fisheries. The national government also needs to investigate various options and make clear proposals for the surveillance of illegal foreign fishing vessels so that PNG does not continue to lose the economic and biological values of marine species.

11. **Monitor the status of the current moratorium on the live reef fish trade.** There is a current national moratorium on live reef fishing. However, trial licenses may be issued after a management plan for the trade is developed and the fishery is deemed viable.

12. **Establish a long-term environmental monitoring program.** Bi-annual surveys by marine biologists, students, and communities are recommended to monitor the status of reef environments and particular species and promote awareness and interest in their conservation.

13. **Continue to promote inter-agency coordination and collaboration between relevant non-government and government institutions.** Many institutions work in isolation from each other and do not benefit from shared experiences, lessons, expertise, and resources. Another recommendation is that the government commit to developing an integrated coastal management strategy that improves inter-agency coordination.

Overview

INTRODUCTION

Conservation International has developed a highly targeted strategy to address and mitigate the degradation of the coastal and marine environments of Milne Bay Province. This strategy underlies a joint marine conservation project between the Milne Bay Provincial Government (MBPG), Conservation International (CI), the United Nations Development Project (UNDP), and the Department of Environment and Conservation (DEC), with additional support from other important national institutions. Assessments of biological, socio-economic, and legislative issues were undertaken, beginning in 1997, partly under a Global Environment Facility (GEF) planning grant. This preliminary work has now resulted in further grants to facilitate the long-term marine conservation and resource management in Milne Bay Province.

CI has played a crucial role in the formation of the Community-Based Coastal and Marine Conservation Program (CMCP). One of its main goals is to conserve a representative sample of globally significant marine biodiversity in Milne Bay by establishing a community-based resource management framework in partnership with all relevant stakeholders. Hopefully this will be achieved by fostering a positive attitude towards marine conservation and near-shore resource management at various governmental levels (including Province, Ward, local, and community). One of the key elements of this approach will be to establish a representative network of community-based marine conservation and sustainable near-shore resource management areas. In addition, an environmental education program and various conservation awareness activities are planned to help reinforce marine conservation values and develop resource management skills in both formal and informal educational settings.

In August 2000, results from the current and previous marine RAP, in conjunction with socio-economic and other pertinent data, were presented at a participatory workshop to determine which areas in the province would be targeted by the project. Three zones were selected and are shown on Map 2 on page 170. Marine Management and Conservation Areas (MMCA) are indicated within each zone. A network of marine management and conservation sites will be established within these areas, which, due to social and economic factors, are favorable for the possible establishment of MPAs.

The 1997 CI Marine RAP report was officially launched by the Governor of Milne Bay Province in conjunction with the stakeholder workshop. This event generated considerable national and local media interest and helped raise awareness of Milne Bay's extraordinary marine environment. The report was also widely distributed throughout the province, resulting in greater appreciation and support of conservation efforts by resource owners and provincial policy-makers. In addition, CI's environmental education program, in collaboration with the National Department of Education (NDOE), has developed a coral reef manual for teachers, which will be distributed for trials in Milne Bay Province primary schools. If successful, the manual will be used throughout PNG.

In the past year, CI has assisted with the formation of a Milne Bay Province tourism working group, consisting of stakeholders from the private sector, communities, and the Tourism Bureau. In addition, CI facilitated a dive ecotourism workshop involving pertinent stakeholders. This resulted in the introduction of a user-fee system whereby scuba divers pay a small fee to landowners for use of their reefs. Local communities now have another tangible incentive for conserving their marine resources.

To demonstrate CI's long-term commitment to marine conservation in Milne Bay Province, the project has opened a field office in Alotau, the provincial capital. A brief summary of major project activities is described in Overview Table 1.

Marine RAP

There is an obvious need to identify areas of global importance for wildlife conservation and management. However, there is often a problem in obtaining the required data, considering that many of the more remote regions are inadequately surveyed. Scarcity of data, in the form of basic taxonomic inventories, is particularly true for tropical ecosystems. Hence, Conservation International has developed a technique for rapid biological assessment. The method essentially involves sending a team of taxonomic experts into the field for a brief period, often 2–4 weeks, in order to obtain an overview of the flora and fauna. Although most surveys to date have involved terrestrial systems, the method is equally applicable for marine and freshwater environments.

One of the main differences in evaluating the conservation potential of terrestrial and tropical marine localities involves the emphasis placed on endemism. Terrestrial conservation initiatives are frequently correlated with a high incidence of endemic species at a particular locality or region. Granted,

other aspects need to be addressed, but endemism is often considered as one of the most important criteria for assessing an area's conservation worth. Indeed, it has become a universal measure for evaluating and comparing conservation "hot spots." In contrast, coral reefs and other tropical marine ecosystems frequently exhibit relatively low levels of endemism. This is particularly true throughout the "coral triangle" (the area including northern Australia, the Malay-Indonesian Archipelago, Philippines, and western Melanesia), considered to be the world's richest area for marine biodiversity. The considerable homogeneity found in tropical inshore communities is in large part due to the pelagic larval stage typical of most organisms. For example reef fish larvae are commonly pelagic for periods ranging from 9 to 100 days (Leis, 1991). A general lack of physical isolating barriers and numerous island "stepping stones" have facilitated the wide dispersal of larvae throughout the Indo-Pacific.

The most important feature to assess in determining the conservation potential of a marine location devoid of significant endemism is overall species richness or biodiversity. Additional data relating to relative abundance are also important. Other factors requiring assessment are more subjective and depend largely on the observer. Obviously, extensive biological survey experience over a broad geographic range yields the best results. This enables the observer to recognize any unique assemblages within the community, unusually high numbers of normally rare taxa, or the presence of any unusual environmental features. Finally, any imminent

Overview Table 1. Coastal and Marine Conservation Project (CMCP) activities to date (Seeto, 2001).

Activity	Description
1. Rapid Assessment Programs	Quickly catalogued the biodiversity in Milne Bay Province and created awareness at all levels on the state of the coral reefs.
2. Site Selection Workshop	Involved all stakeholders to determine general areas/zones that would be targeted by the project.
3. Conservation Needs Assessment	Filled critical gaps in biogeographical data and determined high biodiversity priority areas within Milne Bay Province.
4. Threats Assessment	Outlined current and emergent threats to marine biodiversity in Milne Bay Province, their root causes, and strategies to address these threats.
5. Social Evaluation Study	Filled critical gaps in socio-cultural data necessary for project design and ascertained community interest and capabilities to participate in conservation activities.
6. Stakeholder Participation Plan	Guided the choice of conservation activities, design of interventions, and implementation processes.
7. Policy and Planning Needs Assessment	Determined a toolbox of legislative options from which the best model can be selected specifically to suit the CMCP.
8. Sustainable Use Options Plan	Identified all possible barriers to the success of the project and determined strategies to overcome these barriers.
9. Monitoring and Evaluation Plan	Determined a comprehensive set of performance indicators to monitor all aspects of the project and allow for appropriate changes to be made where necessary throughout the life of the project.

threats such as explosive fishing, use of cyanide, over-fishing, and nearby logging activities need to be considered.

Reef corals, fishes, and molluscs are the primary biodiversity indicator groups used in Marine RAP surveys. Corals provide the major environmental framework for fishes and a host of other organisms. Without reef-building corals, there is limited biodiversity. This is dramatically demonstrated in areas consisting primarily of sand rubble, or mud. Fishes are an excellent survey group as they are the most obvious inhabitants of the reef and account for a large proportion of the reef's overall biomass. Furthermore, fishes depend on a huge variety of plants and invertebrates for their nutrition. Therefore, areas rich in fishes invariably have a wealth of plants and invertebrates. Molluscs represent the largest phylum in the marine environment, the group is relatively well known taxonomically, and they are ecologically and economically important. Mollusc diversity is exceedingly high in the tropical waters of the Indo-Pacific, particularly in coral reef environments. Gosliner *et al.* (1996) estimated that approximately 60 percent of all marine invertebrate species in this extensive region are molluscs. Molluscs are particularly useful as a biodiversity indicator for ecosystems adjacent to reefs where corals are generally absent or scarce (e.g., mud, sand, and rubble bottoms).

One of the recommendations of the 1997 Milne Bay RAP report was that additional surveys be conducted in view of the large area occupied by Milne Bay Province. The 2000 survey is a direct consequence of this recommendation. The two Milne Bay surveys to date are an integral part of CI's coral reef survey program. It was decided at the Marine RAP Workshop in Townsville, Australia (May 1998), that survey activities should be focused on the "Coral Triangle," because it is the world's richest area for coral reef biodiversity and also its most threatened. Accordingly, CI has now completed five surveys over the past four years: (1) Milne Bay I in 1997 (Werner and Allen, 1998), (2) Calamianes Islands, Philippines in 1998 (Werner and Allen, 2001), (3) Togean and Banggai Islands, Indonesia in 1998 (Allen and McKenna, 2001), (4) Milne Bay II in 2000 (this report), and (5) Raja Ampat Islands, Indonesia in 2001 (McKenna et al. 2002).

Physical Environment

Two seasons are experienced annually in Milne Bay Province. Dry southeasterly trade winds blow almost continuously between May and August, and the northwesterly, rain-bearing monsoon prevails from December to March. The province is also affected by cyclones between November and April. These develop in western Melanesia and rarely extend further north than 13° south latitude, and hence only the most southeasterly areas of Papua New Guinea, chiefly the Milne Bay Islands, are affected (McAlpine, Keig and Falls, 1983; McGregor, 1990). Accordingly, extensive cyclone damage is occasionally experienced on barrier reefs of the Louisiade Archipelago. Milne Bay Province has moderately high temperatures with little seasonal variability, and the average annual rainfall ranges between 2,500 and 3000 mm (King and Ranck, 1980). In the Louisiade Archipelago, rainfall is generally heavier in January to May, with June to August being the driest months. At Nuakarta and the Engineer Group, the dry season extends from November to February. Misima Island recorded 3493.3 mm of rain over 256 days, and Brooker Island had 1468.5 mm over a 147-day period from a one-year period from October 1998 to September 1999 (Kinch, 1999).

Large-scale oceanic events such as the El Nino-Southern Oscillation (ENSO) influence coastal marine environments, creating changes in current patterns and causing unseasonal droughts, especially in drier areas of Milne Bay Province such as the Calvados Chain. The 1997 RAP survey found that current patterns were highly complex and strongly localized. The strongest currents were encountered in reef passages and the channels between islands, and were noted on or near Catsarse Reef (site 6), northeastern Dark Hill Point (site 16), Keast Reef (site 19), Marx Reef (site 37), Swinger Opening (site 43), and Hudumuiwa Pass (site 47).

Sea temperatures during the survey ranged from approximately 26°C to 30°C, with the northern section of Milne Bay experiencing warmer temperatures (ranging from 28°C to 30°C), and the southern section experiencing cooler temperatures (ranging from 26°C to 28°C). The highest temperatures were recorded at Fergusson Island, Amphletts Group, and Cape Vogel. The lowest temperatures were encountered in the western part of the Calvados Chain (sites 31–34).

Human Environment

Milne Bay Province has a total population of 205,000, about 75 percent of which live on offshore islands. Subsistence and artisenal activities such as fishing, harvesting of sedentary marine resources, hunting, and agriculture constitute the bulk of the rural economy, as well as remuneration and trade store income. Availability of continuous food resources is an issue facing all communities. Although population density for the whole province is low, the number of people per unit of arable land is relatively high, especially on small islands, resulting in a high dependence on coastal and marine resources (Kinch, 2001).

Most societies in Milne Bay Province are matrilineal. At present, customary rights over sea areas and resources are poorly defined. Conflicts over fisheries resources and stricter enforcement of private fishing areas are consequences of the increasing value of marine products and the decline of certain stocks (Kinch, 2001). Although the complex nature of customary marine tenure can often create obstacles to conservation (e.g., the establishment of MPAs), once these obstacles are overcome then conservation efforts are more likely to be sustainable in the long-term (Seeto, 2001). Provincial government sources estimate that churches provide 65 percent of all rural services in Milne Bay Province, with churches and other non-governmental organizations administering approximately half of the schools in the province. There are currently 181 elementary schools, 176 primary schools, seven secondary schools, and eight vocational schools. There are approximately 48 languages spoken in Milne Bay, with a literacy rate of 77 percent, second only to the National Capital District (Kinch, 2001).

In 1994, Milne Bay was among the top five provinces with regards to malnutrition rates (Department of Health, 1996). Life expectancy is 52.6 yrs for males and 53.6 yrs for females. Infant mortality was estimated at 70 deaths per 10,000 births, and there has been no substantial improvement since 1980 (Hayes and Lasia, 1999). Malaria and pneumonia continue to be the most common causes of mortality; however perinatal death still remains the number one cause of mortality in the province (Kinch, 2001).

SURVEY SITES AND METHODS

General sites were selected by a pre-survey analysis that relied on literature reviews, nautical charts (particularly Australian marine navigation charts Aus 381, 382, 384, 519, 568, 629, and 630), and consultation with Rob Vanderloos and Wayne Thompson, owners and operators of the respective charter diving boats *Chertan* and *Marlin 1*. Detailed site selection was accomplished upon arrival at the general area, and was further influenced by weather and sea conditions.

At each site, the Biological Team conducted underwater assessments that produced species lists for key coral reef indicator groups. General habitat information was also recorded; as was the extent of live coral cover at several depths. The main survey method consisted of direct underwater observations by diving scientists, who recorded species of corals, molluscs, and fishes. Visual transects were the main method for recording fishes and corals in contrast to molluscs, which relied primarily on collecting live animals and shells (most released or discarded after identification). Relatively few specimens were preserved for later study, and these were invariably species that were either too difficult to identify in the field or were undescribed. Further collecting details are provided in the chapters dealing with corals, molluscs, and fishes.

Concurrently, the Reef Resources and Condition Team used a 50-m line transect placed on top of the reef to record substrate details and observations on key indicator species (for assessing fishing pressure) such as groupers and Napoleon Wrasse. Additional information about utilization of marine resources was obtained when and where possible through informal interviews with villagers.

Survey activities were based aboard two live-aboard diving vessels, Chertan (sites 1–28), and Marlin 1 (sites 29–57). Both vessels were fully equipped for diving and provided vital logistic support in the form of air compressors, scuba tanks, crew assistance, and auxiliary dive boats.

Details for individual sites are provided in the reef condition section (Technical Paper 5 in this report). Overview Table 2 provides a summary of sites. Their location is also indicated on Map 1 on page 169.

Detailed results are given in the separate chapters for corals, molluscs, fishes, reef fisheries, reef condition, and com-

Overview Table 2. Summary of survey sites for Marine RAP survey of Milne Bay Province.

No.	Date	Location	Coordinates
1	30/5/2000	Cobb's Cliff, Jackdaw Channel	10°12.682'S, 150.53.067'E
2	30/5/2000	Bavras Reef, off Northwest Normanby Island, D'Entrecasteaux Islands.	09°51.638'S, 150°46.167'E
3	31/5/2000	Sebulgomwa Point, Fergusson Island, D'Entrecasteaux Islands.	09°43.357'S, 150°50.380'E
4	31/5/2000	Scrub Islet, off Sanaroa Island, D'Entrecasteaux Islands	09°39.984'S, 150°59.892'E
5	31/5/2000	Sanaroa Passage, off Cape Doubtful, Fergusson Island	09°37.558'S, 150°56.551'E
6	1/6/2000	Catsaise Reef, Amphlett Group	09°20.939'S, 150°48.612'E
7	1/6/2000	Toiyana Island, Amphlett Group	09°18.909'S, 150°51.100'E
8	1/6/2000	Urasi Island, Amphlett Group	09°13.532'S, 150°52.383'E
9	2/6/2000	Patch reef west of Wamea Island, Amphlett Group	09°13.554'S, 150°47.930'E
10	2/6/2000	Rock islet near Noapoi Island, Amphlett Group	09°18.909'S, 150°51.100'E
11	2/6/2000	East side of Kwatota Island, Amphlett Group	09°18.639S, 150°51.100'E
12	3/6/2000	Northwest corner of Kwatota Island, Amphlett Group	09°17.210'S, 150°42.139'E
13	3/6/2000	Sunday Island, North of Cape Labillardiere, Fergusson Island	09°16.113'S, 150°30.205'E
14	4/6/2000	Off Mukawa Village, North of Cape Vogel	09°37.796'S, 149°58.572'E
15	5/6/2000	Northeast of Baiawa Village, South Collingwood Bay	09°35.150'S, 149°30.672'E
16	5/6/2000	Offshore patch reef Northeast of Dark Hill Point	09°33.880'S, 149°30.672'E

continued

Overview Table 2. Summary of survey sites for Marine RAP survey of Milne Bay Province. *(continued)*

No.	Date	Location	Coordinates
17	5/6/2000	Sidney Islands	09°34.459'S, 149°48.855'E
18	6/6/2000	Ipoteto Island, Kibirisi Point, Cape Vogel	09°38.077'S, 150°00.923'E
19	6/6/2000	Keast Reef, Ward Hunt Strait	09°34.478'S, 150°03.962'E
20	6/6/2000	South of Kibirisi Point, Cape Vogel	09°39.497'S, 150°01.266'E
21	7/6/2000	South of Ragrave Point, Cape Vogel	09°39.497'S, 150°03.737'E
22	7/6/2000	Sibiribiri Point, Cape Vogel	09°43.573'S, 150°03.197'E
23	7/6/2000	Tuasi Island, North Goodenough Bay	09°46.739'S, 149°53.545'E
24	8/6/2000	Pipra Bay, South Goodenough Bay	10°03.975'S, 149°57.220'E
25	8/6/2000	Guanaona Point, South Goodenough Bay	10°04.628'S, 150°03.245'E
26	8/6/2000	Bartle Bay, South Goodenough Bay	10°05.990'S, 150°09.401'E
27	9/6/2000	Kuvira Bay, South Goodenough Bay	10°10.541'S, 150°17.361'E
28	9/6/2000	Awaiama Bay	10°12.531'S, 150°31.907'E
29	12/6/2000	Gabugabutau Island, Conflict Group	10°43.500'S, 151°44.530'E
30	13/6/2000	Tawal Reef, Louisiade Archipelago	11°02.61'S, 152°21.28'E
31	13/6/2000	Ululina Island, Calvados Chain, Louisiade Archipelago	11°04.67'S, 152°31.40'E
32	14/6/2000	Bagaman Island, Calvados Chain, Louisiade Archipelago	11°07.312'S, 152°42.439'E
33	14/6/2000	Yaruman Island, Calvados Chain, Louisiade Archipelago	11°08.15'S, 152°47.95'E
34	14/6/2000	Abaga Gaheia Island, Calvados Chain, Louisiade Archipelago	11°09.36'S, 152°55.04'E
35	15/6/2000	Lagoon patch reef, West of Sabarl Island, Louisiade Archipelago	11°06.93'S, 152°03.25'E
36	15/6/2000	Wanim Island, Calvados Chain, Louisiade Archipelago	11°15.85'S, 153°05.58'E
37	16/6/2000	Marx Reef, North of Tagula (Sudest) Island	11°23.83'S, 153°26.36'E
38	16/6/2000	Passage Northwest of Mt. Ima, Tagula (Sudest) Island	11°26.20'S, 153°26.50'E
39	17/6/2000	Rossel Passage	11°20.38'S, 153°39.05'E
40	17/6/2000	West Point, Rossel Island	11°22.72'S, 153°58.19'E
41	18/6/2000	Patch reef North of Mboibi Point, Rossel Island	11°19.45'S, 154°00.26'E
42	18/6/2000	North side of Wola Island, Rossel Lagoon	11°18.10'S, 154°00.55'E
43	18/6/2000	Swinger Opening, Rossel Lagoon	11°16.53'S, 153°57.40'E
44	19/6/2000	Northwest side of outer barrier reef, Rossel Island	11°15.51'S, 153°40.99'E
45	19/6/2000	West tip of outer barrier reef, near Rossel Passage	11°18.76'S, 153°37.07'E
46	19/6/2000	Osasi Island, North Tagula (Sudest) Island	11°21.68'S, 153°19.75'E
47	20/6/2000	Hudumuiwa Pass, Louisiade Archipelago	11°15.63'S, 153°19.82'E
48	20/6/2000	Siwaiwa Island, Louisiade Archipelago	11°03.47'S, 152°56.37'E
49	21/6/2000	Kei Keia Reef, Louisiade Archipelago	11°06.36'S, 152°15.67'E
50	21/6/2000	Horrara Gowan Reef, Louisiade Archipelago	11°00.75'S, 152°18.55'E
51	22/6/2000	Panasia Island, Louisiade Archipelago	11°00.85'S, 152°19.92'E
52	22/6/2000	Pana Rai Rai Island, Louisiade Archipelago	11°14.38'S, 152°10.29'E
53	22/6/2000	Jomard Entrance, Pana Waipona Island, Louisiade Archipelago	11°15.31'S, 152°08.11'E
54	22/6/2000	Punawan Island, Bramble Haven	11°11.43'S, 152°01.54'E
55	23/6/2000	Bently Island, South Engineer Group	10°43.00'S, 151°13.22'E
56	23/6/2000	North Point, Basilaki Island	10°35.09'S, 151°01.68'E
57	24/6/2000	Negro Head, Sideia Island	10°31.85'S, 150°51.57'E

munity use of marine resources, but the key findings of the survey are summarized here.

Results

Biological diversity

The results of the RAP survey indicate that Milne Bay Province is inhabited by a diverse array of coral reef organisms. Totals for the three major indicator groups (Overview Table 3) surpassed those for previous RAPs in both Indonesia and the Philippines.

Corals

A total of 430 coral species were observed or collected, including more than half of the current world total of 794 zooxanthellate species. Approximately 75 species were found, which at the time of the survey had not been reported from Papua New Guinea in previous publications. Most of these, including about 20 new species, were subsequently described and illustrated by Veron (2000). However, at least 22 species from the current survey remain unreported in published literature. Also, numerous unusual growth forms, many of which may eventually represent additional new species, were noted. Underwater photographs and small samples of every new record and potential new species were obtained. The number of species per site ranged from 44–122. The best sites for overall coral diversity were Gabugabutau Island (Conflict Group), Negro Head (N. Sideia Island), near Siwaiwa Islet (Louisiades), and Keikaeia Reef (Louisiades). These same sites, in addition to North Point on Basilaki Island and Punawan Island at Bramble Haven, also rated highly from an aesthetic point of view (good diversity, pristine condition, extensive cover, and good visibility).

Molluscs

Data were obtained only for the first half of the survey, and therefore exclude the Louisiades, Basilaki, and Sidea. A total of 643 species was recorded with a range of 34–119 per site. Combined with the results of the 1996 survey, the total number of molluscs currently known from Milne Bay Province is 954 species. The most abundant species at each site were generally burrowing arcid bivalves, *Pedum spondyloidaeum*, *Lithophaga* sp., and *Coralliophila neritoidea*.

The coral predator *Drupella* was present at every site, but in relatively low numbers. The best sites for molluscs were Scrub Islet (D'Entrecasteaux Islands), near Noapoi Island and Toiyan Island (Amphlett Group), Kuvira Bay (southern Goodenough Bay), and Sebulgomwa Point (Fergusson Island).

Fishes

A total of 798 species or approximately 72 percent of the known reef fish fauna of Milne Bay Province was recorded. The most speciose families included the Pomacentridae, Labridae, Gobiidae, Serranidae, Apogonidae, and Chaetodontidae. The number of species observed at each site ranged from 140 to 260 with an average of 195 species. The highest site diversity was generally found in the Cape Vogel area (207 species per site). The best individual sites were Scrub Islet, near Sanaroa Island, D'Entrecasteau Group (260 species), Mukawa Bay, Cape Vogel area (245), near Rossel Passage (238), and Gabugabutau Island, Conflict Group (235). The overall total reef fish fauna of Milne Bay resulting from the 1997 and 2000 RAP surveys is 1109 species. One new species belonging to the damselfish (Pomacentridae) genus *Pomacentrus* was collected and photographed. It has also been recorded at other locations in Papua New Guinea, as well as the Solomon Islands and Vanuatu.

Reef fisheries

Edible reef fishes were counted and estimates made of their size on a 30 m-wide path centred along a 100 m transect at two depths (approximately 8 and 16 m) per site. Counts of edible species at each site ranged from 13 to 42 species (average 29) for the deep transects and 14 to 44 (average 32) for the shallow transects. For both transects combined (i.e., the site total), number of species ranged from 28 to 56 (average 45). Numbers of individuals counted ranged from 62 to 3136 (average 435) for the deep transects and 80 to 2739 (average 375) for the shallow transects. For both transects combined, counts ranged from 142 to 5875 (average 801).

A total of 209 species belonging to 27 families were recorded during the survey. Of these families the Caesionidae, Scaridae, Acanthuridae, Lutjanidae, Holocentridae, Serranidae, Nemipteridae, and Mullidae contained the majority of species. Particularly abundant were the caesonids or fusiliers (most notably *Pterocaesio pisang* and *Caesio cuning*). Of the non-caesonid fauna members of the surgeonfish genus *Ctenochaetus* (Acanthuridae) were the most commonly recorded species.

Giant clams and holothurians

Clams and species of beche de mer were assessed at each site by swimming a random path, noting the general abundance and maximum concentration of the various target species. During the survey both clams and beche de mer species were generally seen in low numbers, believed to be a direct reflection of their harvesting for commercial purposes. Most of the clams seen were *Tridacna crocea*, a small species that

Overview Table 3. Summary of Milne Bay fauna recorded during current and past RAP surveys.

Faunal group	No. families	No. genera	No. species
Reef corals	19	77	430
Molluscs	111	290	945
Fishes	93	357	1109

burrows into hard substratum and is not generally used for commercial purposes.

Turtles

Less than 15 turtles were sighted during the survey; however data from Kinch (1999) for Brooker Island in the West Calvados Chain indicate their abundance and important role in sustaining local livelihoods. The peak nesting period, when turtles and their eggs are harvested, is between November and March, which coincides with the time when very little food is produced in local gardens. In one recorded nesting season between October 1998 and June 1999, Brooker people harvested a total of 149 green turtles and 50 hawksbill turtles, as well as eggs from 604 nests belonging to these two species, from 26 different sites, usually small islands (Kinch, 1999).

Sharks

Aside from the Reef Whitetip (*Triaenodon obesus*), sharks were generally absent from most sites with notable exceptions at two locations on the outer reef (sites 44 and 45) surrounding Rossel Island, where the Grey Reef Shark (*Carcharhinus ambyrhynchos*) was common. A large Silvertip Shark (*C. albimarginatus*) was also sighted at one of these same sites. In addition, a 2-meter long Tiger Shark (*Galeocerdo cuvieri*) was seen near Wamea Island in the Amphlett Group.

Reef Condition

Several parameters were used to evaluate reef condition, including the percentage of live coral cover, and damage caused by bleaching, coral predators, pathogens, cyclones, anchors, nets, pollution, freshwater runoff, and siltation. In addition, both subsistence and commercial reef fishing acitivity were evaluated for each site. This information was combined with the fish and coral diversity results to formulate a relative rating (Reef Condition Index) for each site. Several of the more notable findings are briefly summarized in the following paragraphs.

Coral cover—Percentage cover of live hard corals ranged from 13 to 85 percent with a usual average between 30 and 50 percent. The highest coral cover was recorded at Urasi Island, Amplett Group (85 percent for shallow transect) and Swinger Opening, Rossel Island (76 percent for shallow transect). The richest coral cover was generally recorded on the shallow (8 meter depth) transects.

Coral Bleaching—Coral bleaching was present at every site in Collingwood and Goodenough bays, the D'Entrecasteaux Islands, and Amphlett Islands, with extensive bleaching in the Jackdaw Channel area near East Cape. Damage at other sites was not serious and appeared to be very recent (see also Davis, *et al.,* 1997). The worst affected areas included the vicinity of Noapoi Island (Amphlett Group), Mukawa Bay (Cape Vogel area), and reefs in the southern portion of Collingwood Bay. Bleaching was rarely observed in the Louisiades and negligible in extent. There was a definite correlation between warm sea temperatures and the occurrence

of bleaching. The average water temperature in the northern areas affected by bleaching was 28°C compared to 26–27°C for the non-bleached Louisiades.

Coral pathogens and predators—The coral-feeding mollusc *Drupella* was noted at most sites, but damage was not serious. Coral disease was also prevalent at many sites, but never in threatening amounts, although the worst affected area was the Amphlett Islands. Crown-of-thorns starfish, another coral-feeding species, was extremely rare. Fewer than five animals were observed during the entire survey.

Cyclone damage—Wave damage was periodically noted, but was most severe at two sites in the Louisiades: Tawal Reef near Cormorant Channel and Wanim Island, Calvados Chain.

Siltation—Coastal fringing reefs were invariably affected by well-above-average levels of silt, the result of terrestrial runoff. The worst affected areas were the southern edge of Goodenough Bay, around the islands of the Calvados Chain, Tagula Island, and Rossel Island. The RAP survey was conducted during the peak of the southeasterly season, a time of strong winds and rough seas. Also, this time of year coincides with the beginning of the gardening season where villagers are cutting and burning new gardens. This may have contributed to soil runoff at some of the reef sites. Underwater visibility ranged between 3 and 25 m, with an average value of only 10 m. The best visibility and least siltation were generally encountered at outer reef and passage sites.

Eutrophication—Extensive damage was observed on a patch reef in Rossel Lagoon at the mouth of Yonga Bay. The reef was almost entirely smothered by various algae, including large amounts of *Padina*. This damage was possibly caused by freshwater discharge.

Fishing pressure—Fishing activity was judged to be very light at most locations judging from the abundance of large fishes and lack of fishers, although the weather was poor during much of the survey.

CONSERVATION RECOMMENDATIONS

The results of this Marine RAP survey serve to enhance the findings of the 1997 survey that identified Milne Bay Province as a top marine conservation priority (Conservation International, 1998). This major coral reef "wilderness area" is unique in the Indo-Pacific and represents one of the few last remaining marine ecosystems not yet devastated by the severe environmental threats experienced in the rest of the Coral Triangle. Conservation of Milne Bay's globally significant marine biodiversity will depend upon pro-active, community-orientated preventative measures to address the current and potential threats in the province. Some specific recommendations are as follows:

1. **Evaluate and address threats to the marine environment from land-based activities.** While Milne Bay Province is heavily dependent on the marine environment economically and culturally, terrestrial impacts on this environ-

ment must also be addressed to ensure its protection. The effects of sedimentation on reefs from deforestation, agriculture, and mining are of some concern. Any emerging threats from land-based activities should be closely monitored and appropriate actions taken to mitigate any detrimental effects they may have on the marine environment. Watershed protection also needs to be a primary objective of provincial land-use plans.

2. **Establish community-based marine conservation and resource management areas that result in sustainable fisheries management.** There are numerous tools available to fisheries managers and conservation practioners to achieve this. One that could be considered by communities and agencies for Milne Bay Province is the establishment of MPAs. Currently there are no MPAs in Milne Bay Province. The ultimate success of any MPA is dependent upon the recognition of the biological, social, and economic issues relevant to the local communities and their subsequent incorporation into the selection and design process of MPA establishment. Marine protected areas (MPAs) are effective tools for the conservation and management of marine resources.

3. **Conduct more scientific surveys to fill gaps in biological and habitat data.** Further biological surveys and baseline data collection is required to prioritize more specific areas for MPA establishment. Milne Bay Province occupies a vast area, and several RAP surveys would be required to adequately cover all the important reef areas. Locations that remain unsurveyed include the Trobriand Islands, Woodlark Island, Egum Atoll, Misima Island, Sudest Island south Barrier reef, Bramble Haven, Long/Kossman reef, and the southern mainland coast. Surveys on the remaining populations of sea turtles, sharks, dugongs, and other marine mammals such as whales and dolphins are also needed in order to implement effective and appropriate management plans for their conservation.

4. **Continue collaboration with the National Fisheries Authority on rigorous stock assessments of commercially harvested species, and influence the formulation of species management plans.** The over-harvesting of sea cucumbers, giant clam, and shellfish is a serious concern in Milne Bay Province. Rigorous stock assessments are necessary to gauge the current status of these resources and influence the development of appropriate management plans. The development of management plans for other species at risk of commercial extinction should also be encouraged.

5. **Continue to link tourism benefits to the conservation of marine resources.** The dive tourism industry in Milne Bay has enormous potential for growth, and the number of live-a-board dive boats in the province has grown dramatically in the last decade. To achieve the sustainable conservation of marine resources, the benefits from the dive industry should be shared with the communities to provide incentives for conservation. Dive operators should also be encouraged to spend time in the villages and provide tourists with the option of cultural tours and other land-based activities.

6. **Continue to develop and implement an environmental education and awareness program to impart conservation values to students and communities at all levels.** An environmental education and awareness campaign is required to instill conservation values among students and communities in order to generate support for marine conservation efforts in Milne Bay Province. At a formal level, the PNG school curriculum does not adequately cover marine biology and conservation, or fisheries issues. This problem is exacerbated by the lack of appropriate environmental education materials and teaching aides. At an informal level, communities, church groups, women's groups, and youth groups need to be made aware of the status of their marine environment, how their activities affect marine resources, and how threats to these resources can be addressed and mitigated.

7. **Continue participation in the annual PNG Coastal Cleanup campaign.** This campaign was introduced in 1999 as a collaborative effort between CI and the National Capital District Commission (NCDC). This activity increases public awareness of marine issues in PNG, in particular those pertaining to litter and waste (such as plastic bags), and the detrimental effects they are having on the marine environment and marine species.

8. **Assist in the community mapping of resource ownership.** Customary marine tenure gives control and ownership of most near-shore areas, including reefs, to clans. At present customary marine tenure is loosely defined in Milne Bay Province, and this issue needs to be resolved in order to achieve long-term conservation outcomes and avoid conflicts between clans as resources become scarce or as regulations of MPAs are enforced.

9. **Strengthen capacity within the province for effective implementation of the marine conservation program.** The *Organic Law on Provincial Governments and Local-level Governments (1997)* is responsible for the devolution of power from the central government to the provincial and local-level governments. However, the province has very few staff and resources for conservation or the provision of other services. Strengthening the capacity of the Milne Bay Provincial Government is therefore necessary for it to confront the growing environmental pressures and implement an effective marine conservation program.

10. **Enforce existing laws and propose options for surveillance of illegal foreign fishing vessels.** Greater enforce-

ment of regulations outlined by species management plans is required to address the ongoing problem of over-harvesting and to ensure the long-term viability of these fisheries. The national government also needs to investigate various options and make clear proposals for the surveillance of illegal foreign fishing vessels so that PNG does not continue to lose the economic and biological values of marine species, and an effective monitoring and enforcement system must be developed and implemented to protect these globally endangered species.

11. **Monitor the status of the current moratorium on the live reef fish trade.** During the 1997 marine RAP, CI discovered the only known cyanide fishing operation in Milne Bay, which resulted in the subsequent shut down of this live reef fish operation. There is a current national moratorium on live reef fishing. However, trial licenses may be issued after a management plan for the trade is developed and the fishery is deemed viable.

12. **Establish a long-term environmental monitoring program.** Bi-annual surveys by marine biologists, students, and communities are recommended to monitor the status of reef environments and particular species, and promote awareness and interest in their conservation. Data collected will be useful at a local, national, and global level to monitor the effects on the marine environment from both anthropogenic and natural activities.

13. **Continue to promote inter-agency coordination and collaboration between relevant non-government and government institutions.** Collaboration and cooperation between all relevant agencies (non-government, government, and private) is recommended to ensure that all marine conservation issues and problems are adequately addressed. Many institutions work in isolation from each other and do not benefit from shared experiences, lessons, expertise, and resources. Another recommendation is that the government commit to developing an integrated coastal management strategy that improves inter-agency coordination.

REFERENCES

Allen, G. R. and S. A. McKenna (eds.). 2001. A rapid marine biodiversity assessment of the Togean and Banggai Islands, Sulawesi, Indonesia. RAP Bulletin of Biological Assessment. Washington, DC: Conservation International.

Davies, J. M., R. P. Dunne, and B. E. Brown. 1997. Coral bleaching and elevated seawater temperature in Milne Bay Province, Papua New Guinea, 1996. Marine and Freshwater Research 48(6): 513–516.

Gosliner, T. M., D. W. Behrens, and G. C. Williams. 1996. Coral reef animals of the Indo-Pacific. Monterey, California: Sea Challengers.

Hayes, G. and M. Lasia. 1999. The Population of Milne Bay Province: An Overview of the Demographic Situation and its Implications for Development Planning. Unpublished paper prepared for workshop on Population Projections for Development Planning, 1999. Alotau, Milne Bay Province, Papua New Guinea.

Kinch, J. P. 1999. Economics and Environment in Island Melanesia: A General Overview of Resource Use and Livelihoods on Brooker Island in the Calvados Chain of the Louisiade Archipelago, Milne Bay Province, Papua New Guinea. Unpublished Report for Conservation International, Port Moresby, Papua New Guinea.

Kinch, J. P. 2001. A social evaluation study. Unpublished report for the United Nations Milne Bay Community-Based Coastal and Marine Conservation Program, PNG/99/G41, Port Moresby, Papua New Guinea.

King, D. and S. Ranck (eds.). 1980. Papua New Guinea atlas. Geography Department, University of Papua New Guinea.

Leis, J. M. 1991. The pelagic stage of reef fishes: The larval biology of coral reef fishes. In: Sale, P.F. (ed.). The ecology of fishes on coral reefs. San Diego, California: Academic Press. Pp. 183–230.

McAlpine, J., Keig, G. and Falls, R. 1983. Climate of Papua New Guinea. Canberra: Australian National University Press.

McGregor, G. 1990. Possible Consequence of Climate Warming in Papua New Guinea with Implications for the Tropical South West Pacific Area. In Pernetta, J. and Hughes, P. (eds.). Implications of Expected Climatic Changes in the South Pacific Regions: An Overview. Pp: 25–40. UNEP Regional Seas Report and Studies, No. 128. Nairobi: United Nations Environment Project.

McKenna, S. A., G. R. Allen, and S. Suryadi (eds.). 2002. A rapid marine biodiversity assessment of the Raja Ampat Islands, Papua Province, Indonesia. RAP Bulletin of Biological Assessment. Washington, DC: Conservation International.

Seeto, P. 2001. Establishing community-based Marine Protected Areas in Milne Bay Province, Papua New Guinea. Unpublished paper presented to the Tenth Pacific Science Inter-Congress Symposium on Locally Managed Marine Protected Areas. Guam, June 2001.

Veron, J. E. N. 2000. Corals of the World. Volumes 1–3. Townsville: Australian Institute of Marine Science.

Werner, T. B. and G. R. Allen (eds.). 1998. A rapid biodiversity assessment of the coral reefs of Milne Bay Province, Papua New Guinea. RAP Working Papers 11. Washington, DC: Conservation International.

Werner, T. B. and G. R. Allen (eds.). 2001. A rapid marine biodiversity assessment of the Calamianes Islands, Palawan Province, Philippines. RAP Bulletin of Biological Assessment 17. Washington, DC: Conservation International.

Chapter 1

Corals of Milne Bay Province, Papua New Guinea

Douglas Fenner

SUMMARY

- A list of corals was compiled for 57 sites, including 16 sites on the mainland in the Milne Bay area, 5 sites in the D'Entrecasteaux Islands, 7 sites in the Amphlett Island Group, 26 sites in the Louisiade/Conflict Islands, and 3 sites in the vicinity of the Engineer Group and Basilaki-Sideia islands.

- Milne Bay Province has a very diverse coral fauna. A total of 418 scleractinan corals were observed or collected during the present survey, which compares favorably with the total numbers from the Philippines (411) and Indonesia (427) in recently published reports.

- A total of 494 species of stony coral are known from Papua New Guinea, including previous reports. This clearly places it within the area of the highest coral diversity in the world ("Coral Triangle") along with the Philippines and Indonesia.

- *Acropora, Montipora,* and *Porites* were dominant genera on Milne Bay reefs, with 94, 44, and 27 species respectively.

INTRODUCTION

The principle aims of the coral survey were to provide an inventory of coral species and assess various parameters such as coral species richness, coral cover, and presence of rare species. In addition to mainstream reef corals, the survey includes species growing on sand or other soft sediments within and around reefs. The primary group of corals is the zooxanthellate scleractinian corals, which are those containing single-cell algae and which contribute to reef building. Also included are a small number of zooxanthellate non-scleractinian corals that also produce large skeletons contributing to the reef matrix (e.g., *Heliopora, Tubipora,* and *Millepora*), several azooxanthellate scleractinian corals (*Tubastrea, Dendrophyllia,* and *Rhizopsammia*), and a few azooxanthellate non-scleractinian corals (*Distichopora* and *Stylaster*). All produce calcium carbonate skeletons, which contribute to reef building.

METHODS

Corals were surveyed during about 60 hours of scuba diving by D. Fenner at 57 sites to a maximum depth of 42 m. A list of coral species was recorded at each site. The basic method consisted of underwater observations, usually during a single, 70-minute dive at each site. The name of each identified species was indicated on a plastic sheet on which species names were preprinted. A direct descent was made in most cases to the base of the reef, to or beyond the deepest coral visible. The bulk of the dive consisted of a slow ascent along a zigzag path to the shallowest point of the reef or until further swimming was not possible. Sample areas of all habitats encountered were surveyed, including sandy areas, walls, overhangs, slopes, and shallow reef. Areas typically hosting few or no corals, such as seagrass beds and mangroves, were not surveyed. It is estimated that about 50–60 percent of the corals at an individual site can be recorded with this method, due mainly to the time restriction. For example, on this particular RAP survey the author was assisted by E. Turak, who also recorded coral species at sites 1–28 as part of an independent ecological assessment. The combined Fenner-Turak lists generally contained about 50 percent more species per site. Only species recorded by the author are included in the present analysis in order to maintain consistency with past RAP reports. However, the combined Fenner-Turak species lists for sites 1–28 are incorporated in Appendix 1.

Many corals can be confidently identified to species level while diving, but others require microscopic examination of skeletal features. References used to aid the identification process included Best & Suharsono, 1991; Boschma, 1959; Cairns & Zibrowius, 1997; Claereboudt, M., 1990; Dai, 1989; Dai & Lin, 1992; Dineson, 1980; Hodgson, 1985; Hodgson & Ross, 1981; Hoeksema, 1989; Hoeksema & Best, 1991; Hoeksema & Best, 1992; Moll & Best, 1984; Nemenzo, 1986; Nishihira and Veron, 1995; Ogawa & Takamashi, 1993, 1995; Randall & Cheng, 1984: Sheppard & Sheppard, 1991; Veron, 1985, 1990, 2000; Veron & Pichon, 1976, 1980, 1982; Veron, Pichon & Wijman-Best, 1977; Wallace, 1994, 1997a, 1999; and Wallace & Wolstenholme, 1998.

Corals that could not be readily identified in the field were photographed by D. Fenner (Nikonos V with close-up kit and strobe flash unit), and representative samples (1–3 specimens) of the colonies were collected and labeled for later identification using the taxonomic references cited above. Problematical corals were compared with types and other specimens housed at the Australian Institute of Marine Science.

The list of corals recorded from Milne Bay is still incomplete, due to the time restriction of the survey, the highly patchy distribution of corals, and the difficulty of identifying some species under water. Corals are sufficiently difficult to identify that there are significant differences between leading experts for some identifications.

RESULTS

A total of at least 418 species and 71 genera scleractinian stony corals were recorded during the survey (Appendix 1). Nearly all species were illustrated in Veron (2000).

The number of species recorded at each site is presented in Table 1.1.

The richest sites for coral diversity are presented in Table 1.2. In addition, Appendix 1 contains the complete list of species, including sites where they were recorded.

The three sites (55–57) at Bently, Basilaki, and Sideia islands had the highest average number of species per site, followed by the Louisade Archipelago-Conflict Group, mainland, Amphlett Islands, and the D'Entrecasteaux Islands (Table 1.3).

Table 1.1. Number of species observed at each site.

Site	Species	Site	Species	Site	Species
1	70	20	86	39	106
2	58	21	71	40	55
3	79	22	97	41	44
4	78	23	64	42	81
5	61	24	67	43	106
6	59	25	97	44	70
7	57	26	90	45	57
8	63	27	104	46	92
9	64	28	74	47	53
10	51	29	122	48	114
11	91	30	79	49	109
12	91	31	78	50	85
13	56	32	96	51	63
14	72	33	96	52	69
15	80	34	94	53	91
16	84	35	99	54	107
17	80	36	91	55	103
18	100	37	93	56	88
19	87	38	88	57	121

Table 1.2. Sites with the highest diversity.

Site no.	Location	No. spp.
29	Gabagabutau Island, Conflict Group	122
57	Negro Head, Sideia Island	121
48	Siwaiwa Island, Lousiade Archipelago	114
49	Keikaeia Reef, Lousiade Archipelago	109
54	Punawan Island, Bramble Haven	107
43	Swinger Opening, Rossel Lagoon	106
39	Rossel Passage	106
27	Kuvira Bay, South Goodenough Bay	104
55	Bently Island, South Engineer Group	103
18	Ipoeto Island, Kibirisi Point, Cape Vogel	100

An analysis by habitat (Table 1.4) indicates that outer reefs and passages had the greatest number of species, followed by fringing, lagoon, and platform reefs.

General faunal composition

The coral fauna consists mainly of zooxanthellate (algae-containing, reef-building) Scleractinian corals, with 97 percent of the species in this group. In addition there were six azooxanthellate (lacking algae) scleractinians and 12 corals that were non-scleractinians (helioporids, clavulariids, milleporids, and stylasterids). The most speciose genera are listed in Table 1.5. These 11 genera account for about 63 percent of the total observed species.

The order of the most speciose genera is typical of Western Pacific reefs (Table 1.6), with a few minor differences. *Acropora, Montipora*, and *Porites* are invariably the three most speciose, but for other genera there is generally more variation in the order depending on locality.

Species were added to the overall list at a slow but relatively steady rate after about 15 sites, indicating that sufficient sites were surveyed (Figure 1.1).

Zoogeographic affinities of the coral fauna

The reef corals of Milne Bay Province, and Papua New Guinea in general, belong to the overall Indo-Pacific faunal province. A few species span the entire range of the Indo-Pacific, but most have more limited distributions. Papua New Guinea is situated within the central area of greatest global marine biodiversity, referred to as the Coral Triangle. The highest coral diversity appears to occupy an area enclos-

ing the Philippines and central Indonesia, as well as northern and eastern New Guinea. Areas of lower diversity include eastern Australia's Great Barrier Reef, southern New Guinea, and the Ryukyu Islands of southern Japan. The work of Hoeksema (1992) supports the inclusion of northern New Guinea in the richest area, but that of Best et al. (1989) indicates that coral diversity declines in western Indonesia, which should probably be excluded from the central area of diversity.

Coral diversity declines in all directions from the Coral Triangle, with about 80 species in the Izu Islands near Tokyo, 65 species at Lord Howe Island (off southeastern Australia), 45 species at Hawaii, and about 20 species on the Pacific coast of Panama. Species attenuation is significantly less to the west in the Indian Ocean and Red Sea, although this area is still insufficiently studied to provide accurate figures.

Table 1.3. Species richness by geographic area within Milne Bay Province.

Area	No. sites	Avg. spp.
Southern islands (sites 55–57)	3	104
Louisiades/Conflict Group	26	86
Mainland	16	83
Amphlett Islands	7	68
D'Entrecasteaux Islands	5	66

Table 1.4. Species richness by major habitat types.

Reef type	No. sites	Avg. spp.
Outer reef or passage	16	91
Fringing	24	82
Lagoon	11	79
Isolated platform	6	69

Table 1.5. Most speciose genera recorded for Milne Bay sites.

Genus	No. species	%
Acropora	94	22
Montipora	44	10
Porites	27	6
Fungia	16	4
Favia	15	4
Acanthastrea	14	3
Goniopora	13	3
Pavona	12	3
Leptoseris	12	3
Favites	10	2
Astreopora	10	2

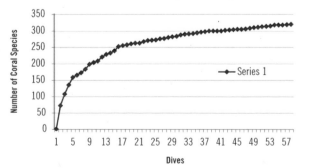

Figure. 1.1. Rate of accumulation of coral species recorded during the RAP survey.

Table 1.6. Genera with the greatest number of species for various western Pacific locations from either previous or present RAP surveys.

Genus	Percentage of fauna						
	Eastern Australia	Western Australia	Philippines	Japan	Calamianes Islands	Togean-Banggai Islands	Milne Bay
Acropora	19	18	17	19	13	16	22
Montipora	9	10	10	9	7	5	10
Porites	5	4	6	6	3	3	6
Favia	4	4	4	4	3	2	4
Goniopora	4	4	3	4	1	1	3
Fungia	4	3	4	3	4	3	4
Pavona	2	3	3	3	4	2	3
Leptoseris	2	2	2	3	4	2	3
Cycloseris	3	2	3	2	1	1	1
Psammocora	2	3	2	2	2	2	2

Corals are habitat-builders and appear to have less niche-specialization than some other groups. They are primarily autotrophic, relying on the products of the photosynthesis of their symbiotic algae, supplemented by plankton caught by filter-feeding and suspension feeding. Most require hard substrate for attachment, but a few grow well on soft substrates. Many corals are found in a relatively wide range of conditions. However, others have restricted habitat requirements with regards to depth, degree of protection from waves and current, exposure to light, and bottom type.

The majority of corals have a pelagic larval stage, with a minimum of a few days of pelagic development for broadcast spawners (most species), and larval settling competency lasting for at least a few weeks. A minority of species release brooded larvae that range from immediate settlement to a long pelagic dispersal period. The main distribution patterns for the Milne Bay Area corals are presented in Table 1.7. Most species have broad distributions that extend well beyond Papua New Guinea, both west and east, and usually also north and south. The remaining species have more limited distributions. *Porites flavus* Veron, 2000 is the only coral species presently known only from Milne Bay Province, although several other potential new endemics are currently being investigated (Veron, pers. comm.).

Species of special interest
Several corals recorded during the survey deserve special attention. Two recently discovered species, *Acropora plumosa* Wallace & Wolstenholme, 1998 and *Acropora cylindrica* Veron and Fenner, in Veron 2000, were found at many sites. The latter has a single corallite at the end of a thick round branch. In addition, many colonies were found with several corallites at the end of branches, and appeared to be intermediate between the *A. cylindrica* and *A. palifera*,

Table 1.7. Distribution patterns of Milne Bay corals.

Distribution	Spp.	%
Extend (usually widely) east & west from PNG	362	85
Restricted to the Coral Triangle	25	6
Extend west, also north and/or south	27	6
Extend west, beyond Coral Triangle	7	1.6
Extend south only	2	0.5
Endemic to PNG	1	0.5

a widespread and common species. *Porites rugosa* Fenner and Veron, in Veron 2000, previously known from one site off eastern Sulawesi, Indonesia, was relatively common, being recorded at 20 sites during the present survey. One of the striking features of the fauna was the large number of *Porites*, *Montipora*, and *Acropora* species that are potentially undescribed. Although the total number of previously described *Montipora* that were identified was typical of Coral Triangle areas, most were found at very few sites.

Approximately 75 species were found, which at the time of the survey had not been reported from Papua New Guinea in previous publications (e.g., Veron, 1993; Table 16). Most of these, including about 20 new species, were subsequently described and illustrated by Veron (2000). However, at least 22 species from the current survey remain unreported in published literature (Table 1.8).

Two coral colonies were found that were exceptionally large for their species. Colonies of *Tubastraea micranthus*

Table 1.8. New records for Papua New Guinea observed, photographed, or collected during the RAP survey.

Pocillopora ankeli	Cycloseris colini
Pocillopora elegans	Pectinia aylini
Montipora cactus	Acanthastrea subechinata
Montipora florida	Lobophyllia serratus
Acropora elegans	Symphyllia hassi
Acropora sekiseinsis	Platygyra acuta
Acropora seriata	Cyphastrea japonica
Porites cumulatus	Rhizopsammia verrilli
Porites negrosensis	Millepora intricata
Coscinaraea monile	Distichopora nitida
Pavona frondifera	Distichopora violacea

about 2 m tall were found at site 39. In additon, a massive 3 m-diameter colony of *Turbinaria stellata* was seen at site 7.

Sites of special interest

A few sites had unusual coral communities of special interest. Perhaps the most outstanding were the following sheltered sites, where the deep reef community consisted of fields of large foliose corals: site 7 (Toiyana Island, Amphlett Islands), site 8 (Urasi Island, Amphlett Islands), site 9 (Patch reef west of Wamer Island, Amphlett Islands), and site 41 (patch reef north of Mboibi Point, Rossel Island). Several species, typically forming a cluster of colonies with large plates growning upward at an angle and arranged in whorls, were represented at each of these sites. *Pachyseris speciosa* was the most common species involved, but others included *Pachyseris foliosa*, *Galaxea astreata*, *Galaxea paucisepta*, *Echinophyllia taylorae*, and an undetermined *Goniopora*. *Echinophyllia taylorae* is usually an encrusting species or forms massive colonies. Moreover, plate-forming colonies of the two *Galaxea* species are not well known. *Acropora pichoni*, a rare table-like species, was also present in one of these communities. Site 36 (Wanim Island, East Calvados Chain, Louisade Archipelago) had a very barren slope, but abundant healthy coral in very shallow water (about 1 m depth or less). An additional three sites were noteworthy due to their outstanding visual appeal: 48 (Siwaiwa Is., Louisade Archipelago), 54 (Punawan Island, Bramble Haven), and 56 (North Point, Basilaki Island).

SUMMARY

The coral fauna of Papua New Guinea is undoubtedly one of the richest in the world. The only other countries with comparable diversity are the Philippines and Indonesia.

The three combined countries form the heart of the Coral Triangle, containing the world's richest coral reefs.

The total scleractinian species (418) documented during the present survey compares favorably with published totals from the Philippines (411 species; Veron and Hodgson, 1989) and Indonesia (427 species; Tomascik, 1997 and Wallace, 1998). However, the coral fauna of both these countries has not been completely documented, and additional species will surely be added. The present species total greatly surpasses that of previous RAP surveys at the Togean-Banggai Islands of Indonesia (301 species) and Calamianes Islands in the Philippines (308), adding further support for Papua New Guinea's importance in the "Coral Triangle." An average of 82 species per site was recorded during the present survey compared with 92.5 species for the Calaminaes and 69.7 species per site in the Togean-Banggai Islands. There were 57 sites in the current survey, 37 sites in the Calamianes, and 47 sites in the Togean-Banggai Islands. Comparing the number of species after 37 sites (the highest common denominator for the three surveys), Milne Bay had 299 species, the Calamines had 304 species, and the Togean-Banggai Islands had 291 species. These data indicate that the sites in Milne Bay were very similar to those in the Philippines and Indonesia for species richness.

Including the present report, at least 481 species of Scleractinia (and a total of 494 species of all stony corals) have been reported from Papua New Guinea. A total count of 498 scleractinians are expected to occur there according to Veron's (2000) distribution maps. Although some species that Veron indicated as occuring at Papua New Guinea are based on projected range extrapolations rather than actual records. In spite of the high diversity documented in the Coral Triangle to date, there is still a need for further study, and no localized area such as Milne Bay Province has been studied comprehensively.

REFERENCES

Best, M. B. and B. W. Hoeksema. 1987. New observations on Scleractinian corals from Indonesia: 1. Free-living species belonging to the Faviina. Zoologische Mcdedelin gen 61: 387–403.

Best, M. B. and Suharsono. 1991. New observations on Scleractinian corals from Indonesia: 3. Species belonging to the Merulinidae with new records of *Merulina* and *Boninastrea*. Zoologische Mededelingen 65: 333–342.

Best, M., B. W. Hoeksema, W. Moka, H. Moll, Suharsono, and I. Nyoman Sutarna. 1989. Recent scleractinian coral species collected during the Snellius-II Expedition in eastern Indonesia. Netherlands Journal of Sea Research 23: 107–115.

Boschma, H. 1959. Revision of the Indo-Pacific species of the genus *Distichopora*. Bijdragen tot de Dierkunde 29: 121–171.

Claereboudt, M. 1990. *Galaxea paucisepta* nom. nov. (for *G. pauciradiata*), rediscovery and redescription of a poorly known scleractinian species (Oculinidae). Galaxea. 9: 1–8.

Claereboudt, M. and J. Bouillon. 1987. Coral associations, distribution and diversity on Laing Island reef (Papua New Guinea). Indo-Malayan Zoology 4: 11–25.

Dai, C-F. 1989. Scleractinia of Taiwan. I. Families Astrocoeniidae and Pocilloporiidae. Acta Oceanographica Taiwan 22: 83–101.

Dai, C-F. and C-H. Lin. 1992. Scleractinia of Taiwan III. Family Agariciidae. Acta Oceanographica Taiwan 28: 80–101.

Dineson, Z. D. 1980. A revision of the coral genus *Leptoseris* (Scleractinia: Fungiina: Agariciidae). Memoirs of the Queensland Museum 20: 181–235.

Hodgson, G. and M. A. Ross. 1981. Unreported scleractinian corals from the Philippines. Proceedings of the Fourth International Coral Reef Symposium 2: 171–175.

Hoeksema, B. W. 1989. Taxonomy, phylogeny and biogeography of mushroom corals (Scleractinia: Fungiidae). Zoologische Verhandelingen 254: 1–295.

Hoeksema, B. W. 1992. The position of northern New Guinea in the center of marine benthic diversity: a reef coral perspective. Proceedings of the Seventh International Coral Reef Symposium 2: 710–717.

Hoeksema, B. W. and M. B. Best. 1991. New observations on scleractinian corals from Indonesia: 2. Sipunculan-associated species belonging to the genera *Heterocyathus* and *Heteropsammia*. Zoologische Mededelingen 65: 221–245.

Hoeksema, B. and C-F. Dai. 1992. Scleractinia of Taiwan. II. Family Fungiidae (including a new species). Bulletin of the Institute of Zoology, Academia Sinica. 30: 201–226.

Moll, H. and M. B. Best. 1984. New scleractinian corals (Anthozoa: Scleractinia) from the Spermonde Archipelago, south Sulawesi, Indonesia. Zoologische Mededelingen 58: 47–58.

Nemenzo, F., Sr. 1986. Guide to Philippine Flora and Fauna: Corals. Manila: Natural Resources Management Center and the University of the Philippines.

Nishihira, M. 1991. Field Guide to Hermatypic Corals of Japan. Tokyo: Tokai University Press.

Nishihira, M. and J. E. N. Veron. 1995. Corals of Japan. Tokyo: Kaiyusha Publishers.

Ogawa, K. and K. Takamashi. 1993. A revision of Japanese ahermatypic corals around the coastal region with guide to identification- I. Genus *Tubastraea*. Nankiseibutu: Nanki Biological Society 35: 95–109 (in Japanese).

Ogawa, K. and K. Takamashi. 1995. A revision of Japanese ahermatypic corals around the coastal region with guide to identification- II. Genus *Dendrophyllia*. Nankiseibutu: Nanki Biological Society 37: 15–33. (in Japanese)

Randall, R. H. and Y-M. Cheng. 1984. Recent corals of Taiwan. Part III. Shallow water Hydrozoan Corals. Acta Oceanographica Taiwan 22: 35–99.

Sheppard, C. R. C. 1998. Corals of the Indian Ocean. CD-ROM. Stockholm: Sida.

Sheppard, C. R. C. and A. L. S. Sheppard. 1991. Corals and coral communities of Arabia. Fauna Saudi Arabia 12: 3–170.

Veron J. E. N. 2000. Corals of the World. Vol. 1-3. Australian Institute of Marine Science.

Veron, J. E. N. 1985. New scleractinia from Australian reefs. Records of the Western Australian Museum 12: 147–183.

Veron, J. E. N. 1986. Corals of Australia and the Indo-Pacific. Univ. Hawaii Press, Honolulu.

Veron, J. E. N. 1990. New scleractinia from Japan and other Indo-West Pacific countries. Galaxea 9: 95–173.

Veron, J. E. N. 1993. A Biogeographic Database of Hermatypic Corals. Australian Institutue of Marine Science Monograph 10: 1–433.

Veron, J. E. N. 1995. Corals in Space and Time. The Biogeography and Evolution of the Scleractinia. Sydney, Australia: University of New South Wales Press.

Veron, J. E. N. 1998. Corals of the Milne Bay Region of Papua New Guinea. *In*: Werner, T. A. and G. R. Allen (eds). A rapid biodiversity assessment of the coral reefs of Milne Bay Province, Papua New Guinea. Washington DC: Conservation International. Pp. 26–34.

Veron, J. E. N. and G. Hodgson. 1989. Annotated checklist of the hermatypic corals of the Philippines. Pacific Science 43: 234–287.

Veron, J. E. N. and M. Pichon. 1976. Scleractinia of Eastern Australia. I. Families Thamnasteriidae, Astrocoeniidae, Pocilloporidae. Australian Institute of Marine Science Monograph Series 1: 1–86.

Veron, J. E. N. and M. Pichon. 1980. Scleractinia of Eastern Australia. III. Families Agariciidae, Siderastreidae, Fungiidae, Oculilnidae, Merulinidae, Mussidae, Pectiniidae, Caryophyllidae, Dendrophyllidae. Australian Institute of Marine Science Monograph Series 4: 1–422.

Veron, J. E. N. and M. Pichon. 1982. Scleractinia of Eastern Australia. IV. Family Poritidae. Australian Institute of Marine Science Monograph Series 5: 1–210.

Veron, J. E. N., M. Pichon, and M. Wijsman-Best. 1977. Scleractinia of Eastern Australia. II. Families Faviidae, Trachyphylliidae. Australian Institute of Marine Science Monograph Series 3: 1–233.

Veron, J. E. N. and C. Wallace. 1984. Scleractinia of Eastern Australia. V. Family Acroporidae. Australian Institute of Marine Science Monograph Series 6: 1–485.

Wallace, C. C. 1994. New species and a new species-group of the coral genus *Acropora* (Scleractinia: Astrocoeniina: Acroporidae) from Indo-Pacific locations. Invert. Taxonomy. 8: 961–88.

Wallace, C. C. 1997a. New species of the coral genus *Acropora* and new records of recently described species from Indonesia. Zool. J. Linn. Soc. 120: 27–50.

Wallace, C. C. 1997b. The Indo-Pacific centre of coral diversity re-examined at species level. Proceedings of the Eighth International Coral Reef Symposium 1: 365–370.

Wallace, C. C. 1999a. The Togian Islands: coral reefs with a unique coral fauna and an hypothesized Tethys Sea signature. Coral Reefs 18: 162.

Wallace, C. C. 1999b. Staghorn corals of the world, a revision of the genus *Acropora*. Collingwood, Australia: CSIRO.

Wallace, C. C. and C.F. Dai. 1997. Scleractinia of Taiwan (IV): Review of the coral genus *Acropora* from Taiwan. Zoological Studies 36: 288–324.

Wallace, C. C. and J. Wolstenholme. 1998. Revision of the coral genus *Acropora* in Indonesia. Zool. J. Linn. Soc. 123: 199–384.

Chapter 2

Condition of coral reefs in Milne Bay Province

Gerald R. Allen, Pamela Seeto, and Tessa McGarry

SUMMARY

- Reef condition is a term pertaining to the general "health" of a particular site as determined by assessment of key variables including natural and human-induced environmental damage and general biodiversity as defined by major indicator groups (corals and fishes).

- Reef condition was assessed at 57 sites in Milne Bay Province, including southern Collingwood Bay, Goodenough Bay, Amphlett Islands, and the Louisiade Archipelago.

- A Reef Condition Index (RCI) value was calculated for each site. Essentially it is a semi-quantitative measure and is derived from three components: coral diversity, fish diversity, and relative damage from human and natural causes. The latter category also incorporates the percentage of live coral cover.

- The hypothetical maximum RCI for a pristine reef is 300. RCI values are useful for interpreting reef condition and comparing sites. Depending on their RCI, sites can be classified as extraordinary, excellent, good, moderate, poor, and very poor. The frequency of Milne Bay sites was as follows: extraordinary (6), excellent (12), good (10), moderate (21), poor (8), and very poor (0).

- The highest RCI value (258.82) was recorded for Gabugabutau Island in the Conflict Group. Major geographic areas with the highest mean RCIs include Basilaki-Sidea (239.55), Conflict-Louisiade Archipelago groups (206.58), and the Cape Vogel area (203.07).

- The mean RCI value for surveyed areas of Milne Bay Province (199.16) was significantly greater than the value (179.87) obtained for the Togean-Banggai Islands of Indonesia.

- Coral bleaching was recorded at every site in Milne Bay north of 10°30' South latitude, and was conspicuously absent at southern sites, including the vast Louisiade Archipelago. The discrepancy is apparently correlated with sea temperatures, which averaged 28.27° and 26.70° for northern and southern areas respectively.

- Although widespread in southern Collingwood Bay, Goodenough Bay, the D'Entrecasteaux Islands, and Amphlett Islands, coral bleaching was not serious, except for one site near East Cape.

- Minor damage due to coral pathogens and the coral-feeding mollusc *Drupella* was also confined mainly to northern reefs. Crown-of-thorns starfish (*Acanthaster planci*), another well-documented predator of scleractinian corals, was rarely seen.

INTRODUCTION

Coral reefs are the most important marine environment in Milne Bay Province. Although there is no precise information on the area occupied, reefs are well developed throughout the province. The extensive but narrow fringing reefs along much of the coastal mainland and islands are among the most obvious structures. Major island groups include the Trobriands, D'Entrecasteauxs, Louisiades, Woodlark and the cluster immediately southeast of Milne Bay proper, including Sariba, Sideia, and Basilaki. There are also many smaller groups such as the Luscany, Amphlett, Egum, Marshall Bennett, Laseinie, Engineer, Conflict, and Bonvouloir. In addition, there is a legion of small reefs and shoals. The northernmost section of the province, lying west of the Trobriands and north of the D'Entrecasteaux Group, is dotted with literally hundreds of such reefs.

The very existence of most Milne Bay islanders is inextricably linked to its bountiful coral reef heritage. Reef resources not only provide food and a means of earning cash (sometimes the only means), they also have strong cultural significance and provide a beautiful home landscape. It is therefore vitally important to assess and monitor the condition or "health" of reefs, and provide guidelines for their conservation and management. Hence, Marine RAP surveys also assess live coral cover and various threats, past or present, that may exert deleterious effects on this fragile ecosystem. This information is combined with biodiversity data from each site to give an approximate indication of reef condition.

MATERIALS AND METHODS

Definition of reef condition

Reef condition is used here as a term that reflects the general "health" of a particular site as determined by an assessment of variables that include environmental damage due to natural and human causes and general biodiversity as defined by key indicator groups (corals and fishes). It also takes into account amounts of live scleractinian coral cover.

Reef Condition Index (RCI)

RAP surveys provide an excellent vehicle for rapid documentation of biodiversity of previously unstudied sites. They also afford an opportunity to issue a "report card" on the status or general condition of each reef site. However, this task is problematical. The main challenge is to devise a rating system that is not overly complex and accurately reflects the true situation, thus providing a useful tool for comparing all sites for a particular RAP or for comparing sites in different regions. CI's Reef Condition Index (RCI) has evolved by trial and error, and although not yet perfected, shows promise of meeting these goals. The present method was trialed during the Togean-Banggai RAP in 1998, and data is now available for 104 sites, including those from the current

survey. Basically, it consists of three components: fish diversity, coral diversity, and condition factors.

Fish diversity component—Total species observed at each site. A hypothetical maximum value of 280 species is utilized to achieve equal weighting. Therefore, the species total from each site is adjusted for equal weighting by multiplying the number of species by 100 and dividing the result by 280.

Coral diversity component—Total species observed at each site. A hypothetical maximum value of 130 species is utilized to achieve equal weighting. The species total from each site is adjusted for equal weighting by multiplying the number of species by 100 and dividing the result by 130.

Reef condition component—This is the most complex part of the RCI formula, and it is therefore instructive to give an example of the data taken from an actual site (site 1):

Parameter	1	2	3	4
1. Explosive/cyanide damage				X
2. Net damage				X
3. Anchor damage				X
4. Cyclone damage				X
5. Pollution/eutrophication			X	
6. Coral bleaching		X		
7. Coral pathogens/predators			X	
8. Freshwater runoff				X
9. Siltation			X	
10. Fishing pressure		X		
11. Coral cover	X			
Bonus/Penalty Points	-20	-10	+10	+20
Totals	-20	-20	+30	+100

Each of the 10 threat parameters and the coral cover category (11) are assigned various bonus or penalty points, utilizing a 4-tier system that reflects relative environmental damage: 1. excessive damage (-20 points), 2. moderate damage (-10 points), 3. light damage (+ 10 points), 4. no damage (+ 20 points). Coral cover is rated according to percentage of live hard coral as determined by 100 m line transects (see below): 1. < 26%, 2. 26–50%, 3. 51 75%, 4. 76–100%. In the example shown here the resultant point total is 90. The maximum possible value of 220 (pristine reef with all parameters rated as category 4) is used to achieve equal weighting. The points total for each site is adjusted for equal weighting by multiplying it by 100 and dividing the result by 220. Therefore, for this example the adjusted figure is 40.9.

Calculation of Reef Condition Index—The sum of the adjusted total for each of the three main components described above. Each component contributes one third of the RCI, with a maximum score of 100 for each. Therefore, the top RCI for a totally pristine reef with maximum fish and coral diversity would be 300. Of course, this situation probably does not exist.

Interpretation of RCI values—The interpretative value of RCI will increase with each passing RAP. Thus far the complete data set contains 104 sites, 47 from the Togean-Banggai Islands and 57 from Milne Bay Province. Table 1 provides a general guide to interpretation, based on the data accumulated thus far.

Table 2.1. Interpretation of RCI values based on 104 sites.

General reef condition	RCI value	Percentage of sites
Extraordinary	>243	4.81
Excellent	214–242	6.73
Good	198–213	28.84
Moderate	170–197	34.61
Poor	141–169	23.08
Very poor	<140	1.92

Coral cover

Data were collected at each site with the use of scuba-diving equipment. The main objective was to record the percentage of live scleractinian coral and other major substrates, including dead coral, rubble, sand, soft corals, sponges, and algae. A 100-m measuring tape was used for substrate assessment in two separate depth zones at most sites, usually 8 and 16 m. However on several occasions there was insufficient depth, and only one transect was done. Substrate type was recorded at 1 m intervals along the tape measure, resulting in direct percentages of the various bottom types for each zone. For the purpose of calculating RCI, the average percentage of coral cover was used if more than one transect per site was involved.

INDIVIDUAL SITE DESCRIPTIONS

1. Cobb's Cliff, Jackdaw Channel

Time: 0930 hours, dive duration 100 minutes; depth range 3–40 m; visibility approximately 10 m; temperature 28°C; slight current; slight turbidity; *site description*: sheltered lagoon type reef with steep slope to deep water with coral growth to depth of about 20 m then mainly sand bottom; dead coral, rubble, and sponge dominant substrata; dominant coral taxa was *Porites* spp. at 0–8 m depth and *Favites* spp. at 9–20 m depth; hard coral cover = 19% at 6–8 m, 16% at 16–17 m; average hard coral cover 17.5%; light

eutrophication, coral pathogens, and siltation; moderate coral bleaching and fishing pressure; important site for subsistence fishing and trade. RCI = 151.90.

2. Bavras Reef, off NW Normanby Island, D'Entrecasteaux Islands

Time: 1530 hours, dive duration 60 minutes; depth range 4–40 m; visibility approximately 10 m; temperature 28°C; slight current; slight turbidity; *site description*: moderately exposed lagoon type reef with gentle and then vertical slope to deep water; hard coral dominant substrata; dominant coral taxa was *Acropora* spp. at 9–20 m depth; hard coral cover = 77% at 16–17 m (only one transect done); average hard coral cover 77%; light coral bleaching and fishing pressure; no clams recorded; important site for subsistence and small-scale commercial fishing. RCI = 196.60

3. Sebulgomwa Point, Fergusson Island, D'Entrecasteaux Islands

Time: 0845 hours, dive duration 60 minutes; depth range 1–40 m; visibility approximately 15 m; temperature 28°C; no current; *site description*: sheltered fringing reef with seagrass and sand patches and gentle slope to deep water with isolated coral heads at approximately 15 m; hard coral dominant substrata; dominant coral taxa was *Acropora* spp. at 0–8 m depth; hard coral cover = 40% at 5–7 m (only one transect done); average hard coral cover 40%; light eutrophication, coral bleaching, siltation, fishing pressure, and predation by *Drupella*; important site for subsistence fishing and trade. RCI = 187.56

4. Scrub Islet, off Sanaroa Island, D'Entrecasteaux Islands

Time: 1239 hours, dive duration 67 minutes; depth range 1–30 m; visibility approximately 5–10 m; temperature 28°C; slight current; severe turbidity; *site description*: sheltered fringing reef with steep slope to deep water with coral growth to depth of about 20 m then mainly sand bottom; hard coral dominant substrata; dominant coral taxa was *Acropora* spp. at 0–8 m depth and *Favites* spp. at 9–20 m depth; hard coral cover = 57% at 7–9 m, 52% at 17–18 m; average hard coral cover 54.5%; moderate siltation; important site for commercial clam fishing but giant clams nearly all fished out. RCI = 233.91.

5. Sanaroa Passage, off Cape Doubtful, Fergusson Island

Time: 1542 hours, dive duration 90 minutes; depth range 5–35 m; visibility approximately 5 m; temperature 29°C; no current; slight turbidity; *site description*: moderately exposed lagoon type reef with steep slope to deep water; hard coral and soft coral dominant substrata; dominant coral taxa was *Acropora* spp. at 0–8 m depth and *Favites* spp. at 9–20 m depth; hard coral cover = 56% at 16–18 m (only one transect done); average hard coral cover 56%; moderate fishing pressure and light predation by *Drupella*; important site for subsistence fishing and trade. RCI = 193.48.

6. Catsarse Reef, Amphlett Group

Time: 0840 hours, dive duration 64 minutes; depth range 3–45 m; visibility approximately 10–15 m; temperature 28°C; strong current; *site description*: moderately exposed lagoon type reef with steep slope to deep water and large areas of coral rubble; hard coral, rubble, and soft coral dominant substrata; dominant soft coral taxa was *Nepthea* spp. at 0–8 m depth and *Sarcophyton* spp. at 9–20 m depth; hard coral cover = 43% at 4–6 m, 30% at 15–16 m; average hard coral cover 36.5%; light coral bleaching, coral pathogens (white band disease), fishing pressure, and predation by *Drupella*; important site for subsistence fishing and trade. RCI = 198.11.

7. Toiyana Island, Amphlett Group

Time: 1134 hours, dive duration 76 minutes; depth range 1–35 m; visibility approximately 20 m; temperature 29°C; no current; *site description*: exposed fringing reef with moderate slope to deep water and sand channels parallel to slope; sand, hard coral and sponge dominant substrata; dominant coral taxa was *Favites* spp. at 0–8 m and 9–20 m depths; hard coral cover = 32% at 7–9 m, 28% at 16–18 m; average hard coral cover 30%; light eutrophication, coral pathogens (white band disease), predation by *Drupella*, and siltation. RCI = 158.17.

8. Urasi Island, Amphlett Group

Time: 1542 hours, dive duration 75 minutes; depth range 1–30 m; visibility approximately 15 m; temperature 28°C; no current; slight turbidity; *site description*: exposed fringing reef with moderate and then steep slope to deep water; hard coral dominant substrata; dominant coral taxa was *Favites* spp. at 0–8 m depth and *Pocillopora* spp. at 9–20 m depth; hard coral cover = 85% at 6–8 m, 59% at 15–17 m; average hard coral cover 72%; light eutrophication, coral pathogens (white band disease), coral bleaching, and predation by *Drupella*; moderate siltation and fishing pressure; important site for subsistence fishing, trade, and beche-de-mer harvesting. RCI = 170.51.

9. Patch Reef west of Wamea Island, Amphlett Group

Time: 0750 hours, dive duration 74 minutes; depth range 5–40 m; visibility approximately 5 m; temperature 28°C; no current; *site description*: exposed patch reef with moderately steep slope to deep water; hard coral, soft coral, and rubble dominant substrata; dominant coral taxa was *Pocillopora* spp. at 0–8 m depth and *Favites* spp. at 9–20 m depth; hard coral cover = 34% at 8–10 m, 43% at 16–18 m; average hard coral cover 38.5%; light cyclone damage, fishing pressure, and siltation; important site for beche-de-mer harvesting. RCI = 191.24.

10. Rock Islet near Noapoi Island, Amphlett Group

Time: 1021 hours, dive duration 71 minutes; depth range 1–40 m; visibility approximately 10 m; temperature 28°C; no current; slight turbidity; *site description*: exposed fringing reef with steep slope to deep water and sand gullies parallel to slope; hard coral, sand, and rubble dominant substrata; dominant coral taxa was *Favites* spp. at 0–8 m and 9–20 m depths; hard coral cover = 49% at 6–8 m, 46% at 16–17 m; average hard coral cover 47.5%; predation by *Drupella* and light eutrophication; moderate siltation and fishing pressure; important site for beche-de-mer harvesting. RCI = 164.59.

11. East side of Kwatota Island, Amphlett Group

Time: 1448 hours, dive duration 72 minutes; depth range 1–40 m; visibility approximately 10 m; temperature 28–29°C; slight current; slight turbidity; *site description*: sheltered fringing reef of small island with moderate slope to deep water; hard coral dominant substrata; dominant coral taxa was *Favites* spp. at 0–8 m and 9–20 m depths; hard coral cover = 52% at 7–8 m, 60% at 16–17 m; average hard coral cover 56%; light coral pathogens and siltation; moderate fishing pressure; important site for clam harvesting. RCI = 209.87.

12. NW corner of Kwatota Island, Amphlett Group

Time: 0901 hours, dive duration 73 minutes; depth range 1–40 m; visibility approximately 10 m; temperature 28°C; no current; *site description*: moderately exposed fringing reef with gentle to moderate slope to deep water with sandy patches interspersed with coral bommies in the shallows and sand gullies on the slope; hard coral dominant substrata; dominant coral taxa was *Acropora* spp. at 0–8 m and 9–20 m depths; hard coral cover = 61% at 8–9 m, 56% at 15–16 m; average hard coral cover 58.5%; light coral pathogens, fishing pressure, and predation by *Drupella* and *Acanthaster planci* (one individual observed). RCI = 233.51.

13. Sunday Island, north of Cape Labillardiere, Fergusson Island

Time: 1238 hours, dive duration 75 minutes; depth range 3–40 m; visibility approximately 7–10 m; temperature 29°C; no current; slight turbidity; *site description*: exposed fringing reef with moderate slope to deep water; hard coral dominant substrata; dominant coral taxa was *Acropora* spp. at 0–8 m depth and *Favites* spp. at 9–20 m depth; hard coral cover = 65% at 5–6 m, 68% at 15–17 m; average hard coral cover 66.5%; light eutrophication, coral bleaching, fishing pressure, and predation by *Drupella*; moderate coral pathogens and siltation; important site for subsistence fishing, trade, and harvesting of clams and beche-de-mer. RCI = 161.85.

14. Off Mukawa Village, north of Cape Vogel

Time: 1059 hours, dive duration 80 minutes; depth range 3–25 m; visibility approximately 10 m; temperature 29°C; no current; *site description*: sheltered fringing reef with gentle slope to deeper water and patches of reef interspersed with sandy areas; hard coral dominant substrata; dominant coral taxa was *Acropora* spp. at 0–8 m depth and *Porites* spp. at 9–20 m depth; hard coral cover = 39% at 6–8 m (only one transect done); average hard coral cover 39%; predation by *Drupella* and light siltation; moderate eutrophication, coral bleaching, and fishing pressure; important site for local fishermen. RCI = 179.25.

15. NE of Baiawa Village, southern Collingwood Bay

Time: 0750 hours, dive duration 65 minutes; depth range 3–30 m; visibility approximately 5 m; temperature 29°C; no current; *site description*: sheltered fringing reef with gentle slope to deep water; hard coral dominant substrata; dominant coral taxa was *Favites* spp. at 0–8 m depth and *Agariciidae* at 9–20 m depth; hard coral cover = 47% at 6–7 m, 43% at 15–17 m; average hard coral cover 45%; light eutrophication; moderate fishing pressure; heavy siltation; many giant clams and various holothurian species present; very important site for local and commercial fishing. RCI = 180.82.

16. Offshore patch reef NE of Dark Hill Point, southern Collingwood Bay

Time: 1129 hours, dive duration 70 minutes; depth range 3–35 m; visibility approximately 20 m; temperature 29°C; strong current; *site description*: moderately exposed lagoon type reef with moderate slope to deep water; hard coral, rubble, and soft coral dominant substrata; dominant hard coral taxa was *Acropora* spp. at 0–8 m depth and *Fungiidae* at 9–20 m depth; hard coral cover = 52% at 7–9 m, 44% at 16–18 m; average hard coral cover 48%; very light predation by *Drupella*; light to moderate coral bleaching; moderate fishing pressure; important site for beche-de-mer harvesting. RCI = 178.35.

17. Sidney Islands, southern Collingwood Bay

Time: 1420 hours, dive duration 60 minutes; depth range 3–35 m; visibility approximately 15 m; temperature 28° ; no current; *site description*: moderately exposed lagoon type reef with moderate slope to deep water; hard coral and sponge dominant substrata; dominant coral taxa was *Acropora* spp. at 0–8 m and 9–20 m depths; hard coral cover = 66% at 7–9 m, 35% at 15–17 m; average hard coral cover 50.5%; light eutrophication, coral bleaching, coral pathogens, and fishing pressure; important site for beche-de-mer harvesting; one *Acanthaster planci* observed. RCI = 202.03.

18. Ipoteto Island, Kibirisi Point, Cape Vogel

Time: 0745 hours, dive duration 68 minutes; depth range 2–35 m; visibility approximately 25 m; temperature 28°; no current; *site description*: sheltered fringing reef with steep slope to deep water and sand gullies parallel to slope; hard coral, soft coral, and sponge dominant substrata; dominant coral taxa was *Acropora* spp. at 0–8 m depth and *Favites* spp. at 9–20 m depth; hard coral cover = 48% at 7–9 m, 33% at 16–18 m; average hard coral cover 40.5%; light siltation; heavy fishing pressure. RCI = 206.99.

19. Keast Reef, Ward Hunt Strait

Time: 1150 hours, dive duration 65 minutes; depth range 4–50 m; visibility approximately 15 m; temperature 28°C; strong current; *site description*: exposed lagoon type reef with very steep slope to deep water greater than 80 m; hard coral and rubble dominant substrata; dominant coral taxa was *Acropora* spp. at 0–8 m depth and *Favites* spp. at 9–20 m depth; hard coral cover = 70% at 7–8 m, 57% at 16–17 m; average hard coral cover 63.5%; light net damage; moderate fishing pressure; important site for subsistence and commercial fishing. RCI = 218.48.

20. South of Kibirisi Point, Cape Vogel

Time: 1505 hours, dive duration 69 minutes; depth range 2–51 m; visibility approximately 10 m; temperature 28°C; no current; moderate turbidity; *site description*: sheltered fringing reef with very steep slope to deep water and many sandy/rubble areas on slope; hard coral, sponge, and rubble dominant substrata; dominant coral taxa was *Porites* spp. at 0–8 m and 9–20 m depths; hard coral cover = 42% at 7–8 m, 32% at 16–18 m; average hard coral cover 37%; light eutrophication and siltation; important site for subsistence fishing and trade. RCI = 204.69.

21. South of Ragrave Point, Cape Vogel

Time: 0820 hours, dive duration 60 minutes; depth range 1–20 m; visibility approximately 5 m; temperature 28°C; no current; moderate turbidity; *site description*: sheltered (on inside) lagoon/barrier type reef with gentle slope to a depth of 14 m and then mainly sand bottom; slope interspersed with large coral bommies; hard coral and sand dominant substrata; dominant coral taxa was *Acropora* spp. at 0–8 m depth; hard coral cover = 30% at 5–7 m (only one transect done); average hard coral cover 30%; light fishing pressure; important site for beche-de-mer harvesting. RCI = 202.50.

22. Sibiribiri Point, Cape Vogel

Time: 1035 hours, dive duration 62 minutes; depth range 1–35 m; visibility approximately 15 m; temperature 29–30°C; slight current and wave action; *site description*: sheltered fringing reef with moderate and then very steep slope to deep water greater than 40 m; hard coral, sponge, and sand dominant substrata; dominant coral taxa was *Acropora* spp. at 0–8 m depth and *Porites* spp. at 9–20 m depth; hard coral cover = 37% at 6–8 m, 38% at 17–19 m; average hard coral cover 37.5%; light coral bleaching and moderate fishing pressure; holothurians observed at this site. RCI = 220.65.

23. Tuasi Island, northern Goodenough Bay

Time: 1425 hours, dive duration 70 minutes; depth range 1–28 m; visibility approximately 3 m; temperature 28–29°C; no current; *site description*: sheltered fringing reef with gentle slope to deep water with isolated coral bommies and extensive areas of sand; sponge and hard coral dominant substrata; dominant coral taxa was *Favites* spp. at 0–8 m depth and *Agariciidae* at 9–20 m depth; hard coral cover = 24% at 8–9 m, 3% at 16–18 m; average hard coral cover 13.5%; light fishing pressure; heavy siltation; important site for beche-de-mer harvesting. RCI = 149.23.

24. Pipra Bay, southern Goodenough Bay

Time: 0839 hours, dive duration 85 minutes; depth range 0–40 m; visibility approximately 13 m; temperature 28°C; no current; *site description:* sheltered fringing reef with vertical slope to deep water and sand and rubble gullies parallel to slope; hard coral and sponge dominant substrata; dominant coral taxa was *Acropora* spp. at 0–8 m depth and *Favites* spp. at 9–20 m depth; hard coral cover = 54% at 8–9 m, 35% at 16–18 m; average hard coral cover 44.5%; light coral bleaching and freshwater run-off; moderate fishing pressure; heavy siltation. RCI = 159.99.

25. Guanaona Point, southern Goodenough Bay

Time: 1112 hours, dive duration 70 minutes; depth range 1–38 m; visibility approximately 5 m; temperature 28°C; no current; *site description:* sheltered fringing reef with vertical slope to deep water and with sand/rubble patches interspersed on slope; hard coral dominant substrata; dominant coral taxa was *Acropora* spp. at 0–8 m depth and *Agariciidae* at 9–20 m depth; hard coral cover = 57% at 7–9 m, 51% at 17–18 m; average hard coral cover 54%; moderate fishing pressure; heavy siltation. RCI = 196.11.

26. Bartle Bay, southern Goodenough Bay

Time: 1435 hours, dive duration 67 minutes; depth range 1–35 m; visibility approximately 7 m; temperature 28°C; no current; *site description:* sheltered fringing reef with moderate slope to deep water with scattered coral bommies and large gullies of muddy sand; sand, hard coral, and sponge dominant substrata; dominant coral taxa was *Favites* spp. at 0–8 m depth and *Acropora* spp. at 9–20 m depth; hard coral cover = 36% at 7–8 m, 24% at 16–18 m; average hard coral cover 30%; light eutrophication and coral beaching; moderate freshwater run-off and fishing pressure; heavy siltation. RCI = 178.19.

27. Kuvira Bay, southern Goodenough Bay

Time: 1223 hours, dive duration 60 minutes; depth range 1–35 m; visibility approximately 7 m; temperature 28°C; no current; *site description:* sheltered fringing reef with moderate slope to deep water; hard coral dominant substrata; dominant coral taxa was *Porites* spp. at 0–8 m depth and *Acropora* spp. at 9–20 m depth; hard coral cover = 77% at 7–8 m, 30% at 16–17 m; average hard coral cover 53.5%; moderate coral bleaching; heavy freshwater run-off and siltation; important site for subsistence fishing. RCI = 179.58.

28. Awaiama Bay, southern Goodenough Bay

Time: 1701 hours, dive duration 67 minutes; depth range 1–35 m; visibility approximately 5 m; temperature 28–29°C; no current; *site description:* sheltered fringing reef with gentle slope to deep water with small scattered sandy patches and large areas of rubble; hard coral dominant substrata; dominant coral taxa was *Acropora* spp. at 0–8 m depth and *Tubastraea* spp. at 9–20 m depth; hard coral cover = 72% at 7–9 m, 71% at 16–18 m; average hard coral cover 71.5%; light eutrophication, freshwater run-off and siltation; moderate coral bleaching. RCI = 176.73.

29. Gabugabutau Island, Conflict Group

Time: 1152 hours, dive duration 63 minutes; depth range 1–46 m; visibility approximately 25 m; temperature 27°C; slight current; *site description:* exposed outer reef with moderate slope to deep water; hard coral and soft coral dominant substrata; dominant coral taxa was *Porites* spp. at 0–8 m depth with mixed taxa at 9–20 m depth; hard coral cover = 54% at 7–9 m, 54% at 16–18 m; average hard coral cover 54%; light fishing pressure and very light siltation and predation by *Drupella*; important site for local fishermen especially for harvesting turtles. RCI = 258.82.

30. Tawal Reef, Louisiade Archipelago

Time: 1136 hours, dive duration 60 minutes; depth range 2–50 m; visibility approximately 15 m; temperature 27°C; no current; *site description:* exposed barrier type reef with moderate slope to deep water; hard coral and rubble dominant substrata; dominant taxa was soft coral species at 0–8 m depth; hard coral cover = 34% at 6–8 m, 39% at 15–16 m; average hard coral cover 36.5%; light fishing pressure; heavy cyclone damage; very important site for subsistence fishing and trade. RCI = 206.49.

31. Ululina Island, Calvados Chain, Louisiade Archipelago

Time: 1527 hours, dive duration 60 minutes; depth range 1–38 m; visibility approximately 10 m; temperature 27°C; no current; *site description:* moderately exposed lagoon type reef with moderate slope to deep water and wide sand gullies parallel to slope; sand, hard coral, and rubble dominant substrata; dominant taxa was soft coral species at 0–8 m depth and *Stylaster* spp. at 9–20 m depth; hard coral cover = 22% at 7–9 m, 22% at 15–16 m; average hard coral cover 22%; light siltation and fishing pressure; Indo-pacific coral species that have not previously been found in PNG were recorded at this site; white-tip and black-tip reef sharks seen; important site for subsistence fishing. RCI = 204.51.

32. Bagaman Island, Calvados Chain, Louisiade Archipelago

Time: 0942 hours, dive duration 60 minutes; depth range 2–30 m; visibility approximately 5 m; temperature 27°C; slight current; severe turbidity; *site description:* sheltered fringing reef within Calvados lagoon with gentle slope to deep water with large and small coral rock boulders interspersed with sandy and coral patches; hard coral, sand, and rubble dominant substrata; hard coral cover = 38% at 7–8 m, 47% at 15–17 m; average hard coral cover 42.5%; light fishing pressure; moderate siltation; important site for subsistence fishing and diving for commercially valuable resources. RCI = 209.12.

33. Yaruman Island, Calvados Chain, Louisiade Archipelago

Time: 1244 hours, dive duration 60 minutes; depth range 1–28 m; visibility approximately 5–7 m; temperature 27°C; slight current from two directions; severe turbidity; *site description*: moderately exposed fringing reef within lagoon with moderate slope to deep water and coral bommies interspersed between large areas of sand and rubble; hard coral and rubble dominant substrata; dominant coral taxa was *Porites* spp. at 0–8 m and 9–20 m depths; hard coral cover = 33% at 7–8 m, 46% at 15–17 m; average hard coral cover 39.5%; light fishing net and line damage and fishing pressure; moderate cyclone damage and siltation; important site for subsistence fishing and diving for commercially valuable resources. RCI = 184.20.

34. Abaga Gaheia Island, Calvados Chain, Louisiade Archipelago

Time: 1554 hours, dive duration 60 minutes; depth range 1–35 m; visibility approximately 7 m; temperature 27°C; no current; slight turbidity; *site description*: sheltered fringing reef with gentle slope to deep water with large areas of sand interspersed with coral bommies; hard coral, sand, and sponge dominant substrata; hard coral cover = 53% at 6–8 m, 32% at 14–16 m; average hard coral cover 42.5%; light eutrophication and fishing pressure; moderate siltation; important site for subsistence fishing and diving for commercially valuable resources. RCI = 190.17.

35. Lagoon patch reef west of Sabari Island, Louisiade Archipelago

Time: 1032 hours, dive duration 65 minutes; depth range 1–30 m; visibility approximately 15 m; temperature 28°C; no current; *site description*: sheltered patch reef within lagoon with very shallow reef flat and very steep slopes to deep water and then mainly sand bottom; large patches of coral rubble and small sand gullies on slopes; hard coral, sand, and rubble dominant substrata; dominant coral taxa was soft corals at 0–8 m depth; hard coral cover = 48% at 7–8 m, 30% at 16–18 m; average hard coral cover 39%; light siltation; moderate cyclone damage; important site for subsistence fishing and diving for commercially valuable resources. RCI = 214.48.

36. Wanim Island, Calvados Chain, Louisiade Archipelago

Time: 1400 hours, dive duration 60 minutes; depth range 1–30 m; visibility approximately 5 m; temperature 27°C; no current; *site description*: sheltered fringing reef of small island with moderate slope to deep water and extensive areas of rubble; hard coral dominant substrata; dominant coral taxa was *Porites* spp. at 0–8 m depth and mainly rubble at 9–20 m depth; hard coral cover = 20% at 7–8 m, 27% at 15–17 m; average hard coral cover 23.5%; moderate siltation; severe cyclone damage; important site for subsistence fishing and diving for commercially valuable resources; two *Acanthaster planci* observed. RCI = 172.69.

37. Marx Reef, north of Tagula (Sudest) Island, Louisiade Archipelago

Time: 1101 hours, dive duration 75 minutes; depth range 2–44 m; visibility approximately 20 m; temperature 28°C; very strong current over reef top; *site description*: moderately exposed patch reef with very steep slope to deep water and shallow sand channels leading into the reef lagoon; hard coral dominant substrata; hard coral cover = 67% at 6–7 m, 53% at 15–16 m; average hard coral cover 60%; light siltation and fishing pressure; important site for beche-de-mer, *Trochus* spp. and subsistence fishing; several grey reef sharks and one spotted eagle ray seen. RCI = 231.12.

38. Passage northwest of Mt. Ima, Tagula (Sudest) Island, Louisiade Archipelago

Time: 1418 hours, dive duration 60 minutes; depth range 1–40 m; visibility approximately 5 m; temperature 27°C; no current; slight turbidity; *site description*: sheltered patch reef with vertical slope to deep water and with silty sand gullies on slope; sand and hard coral dominant substrata; hard coral cover = 29% at 6–8 m, 25% at 15–17 m; average hard coral cover 27%; severe siltation; important site for subsistence fishing.

39. Rossel Passage, Louisiade Archipelago

Time: 0952 hours, dive duration 70 minutes; depth range 2–45 m; visibility approximately 15 m; temperature 27°C; moderate current; *site description*: exposed barrier type reef around small patch reef in channel, with moderate slope to deep water and wide sand channels with coral bommies and areas of hard rock substrate; hard coral and sand dominant substrata; hard coral cover = 56% at 5–7 m, 43% at 15–17 m; average hard coral cover 49.5%; light fishing pressure; site remote so not used much by local fishermen. RCI = 252.45.

40. West Point, Rossel Island, Louisiade Archipelago

Time: 1455 hours, dive duration 45 minutes; depth range 1–28 m; visibility approximately 3 m; temperature 27°C; no current; slight turbidity; *site description*: moderately exposed patch reef with gentle slope to deep water with sand and scattered coral bommies in the shallows, followed by rock and then mainly sandy bottom; algae and hard coral dominant substrata; hard coral cover = 33% at 6–7 m (only one transect done); light fishing pressure; severe siltation; important site for subsistence fishing. RCI = 163.03.

41. Patch reef north of Mboibi Point, Rossel Island, Louisiade Archipelago

Time: 0811 hours, dive duration 61 minutes; depth range 1–35 m; visibility approximately 5 m; temperature 27°C; no current; slight turbidity; *site description*: sheltered patch reef with very steep slope to deep water with coral growth to depth of about 30 m then mainly sand bottom; also patches of course sand composed of broken shell fragments on slope; algae (*Halimeda*), rubble, and hard coral dominant substrata;

hard coral cover = 22% at 6–7 m, 13% at 15–16 m; average hard coral cover 17.5%; light fishing pressure; moderate cyclone damage and siltation; severe eutrophication; important site for subsistence fishing. RCI = 143.29.

42. North side of Wola Island, Rossel lagoon, Louisiade Archipelago

Time: 1035 hours, dive duration 60 minutes; depth range 1–28 m; visibility approximately 5 m; temperature 27°C; no current; *site description*: sheltered fringing reef with gentle slope to deep water with sand and coral bommies in the shallows and wide sand gullies on the slope; sand and hard coral dominant substrata; dominant hard coral taxa was *Porites* spp. at 0–8 m depth and *Pavona* spp. at 9–20 m depth; hard coral cover = 41% at 7–8 m, 25% at 16–18 m; average hard coral cover 17.5%; severe eutrophication and siltation; important site for subsistence fishing. RCI = 173.93.

43. Swinger Opening, Rossel Lagoon, Louisiade Archipelago

Time: 1403 hours, dive duration 56 minutes; depth range 1–25 m; visibility approximately 20 m; temperature 28°C; strong current; *site description*: exposed barrier reef with steep slope to deep water; hard coral dominant substrata; hard coral cover = 76% at 7–9 m, 36% at 17–18 m; average hard coral cover 55.5%; light coral bleaching; reef not easily accessible to local fishermen; black-tip and white-tip reef sharks seen. RCI = 243.71.

44. Northwest side of outer barrier reef, Rossel Island, Louisiade Archipelago

Time: 0843 hours, dive duration 57 minutes; depth range 2–42 m; visibility approximately 20 m; temperature 28°C; slight current; *site description*: exposed barrier type reef with vertical slope to deep water with caves, gullies, and passages through to the lagoon; algae (*Halimeda*) dominant substrata; hard coral cover = 7% at 14–16 m (only one transect done); severe eutrophication; several white-tip and grey reef sharks seen. RCI = 217.81.

45. West tip of outer barrier reef, near Rossel Passage, Louisiadc Archipelago

Time: 1130 hours, dive duration 74 minutes; depth range 2–43 m; visibility approximately 20 m; temperature 28°C; slight current; *site description*: exposed barrier type reef with gentle slope to deep water and with sand channels parallel to reef; rubble, algae, and hard coral dominant substrata; hard coral cover = 24% at 8–9 m, 15% at 14–16 m; average hard coral cover 19.5%; light coral bleaching. RCI = 206.12.

46. Osasi Island, northern Tagula (Sudest) Island, Louisiade Archipelago

Time: 1608 hours, dive duration 60 minutes; depth range 2–25 m; visibility approximately 3–5 m; temperature 27°C; no current; severe turbidity; *site description*: sheltered lagoon type reef with moderate slope to deep water and then sand, and with sandy areas interspersed with large coral bommies

in the shallows; hard coral and soft coral dominant substrata; hard coral cover = 52% at 6–7 m, 37% at 14–15 m; average hard coral cover 44.5%; severe siltation; important site for subsistence fishing and diving for commercially valuable resources. RCI = 191.29.

47. Hudumuiwa Pass, Louisiade Archipelago

Time: 0811 hours, dive duration 60 minutes; depth range 1–25 m; visibility approximately 15 m; temperature 28°C; strong current through passage; *site description*: exposed barrier type reef with gentle slope down to deep water and sand, then slopes back up, forming a wide u-shaped passage in the middle; sandy substrate with small broken rock fragments and occasional small coral bommies on top of one side of the passage, while the other side has coral reef; sand, hard coral, and sponge are the dominant substrata; hard coral cover = 52% at 4–6 m, 14% at 14–16 m; average hard coral cover 33%; light predation by *Drupella*; important site for subsistence fishing and diving for commercially valuable resources; green turtle and banded sea snake seen, also grey reef and white-tip sharks present. RCI = 164.05.

48. Siwaiwa Island, Louisiade Archipelago

Time: 1344 hours, dive duration 71 minutes; depth range 3–20 m; visibility approximately 10 m; temperature 28°C; no current; *site description*: sheltered lagoon type reef with moderate and then gentle slope to deep water with sand channels horizontal to slope; soft coral and hard coral dominant substrata; dominant taxa was the hard coral, *Porites* spp. and soft corals at 0–8 m depth; hard coral cover = 25% at 6–8 m, 42% at 13–14 m; average hard coral cover 33.5%; light fishing pressure; important site for subsistence fishing and diving for commercially valuable resources. RCI = 244.51.

49. Kei Keia Reef, Louisiade Archipelago

Time: 1135 hours, dive duration 49 minutes; depth range 3–45 m; visibility approximately 10 m; temperature 27° ; no current; *site description*: exposed barrier type reef with moderate slope to deep water with wide sand channels parallel to slope; sand and hard coral dominant substrata; hard coral cover = 28% at 6–8 m, 19% at 16–18 m; average hard coral cover 23.5%; important site for subsistence fishing, diving for commercially valuable resources, and trade. RCI = 235.71.

50. Horrara Gowan Reef, Louisiade Archipelago

Time: 1428 hours, dive duration 66 minutes; depth range 3–40 m; visibility approximately 10 m; temperature 27°C; no current; *site description*: exposed barrier type reef with gentle slope to deep water with scattered coral bommies in the shallows and then vast areas of rubble overgrown with coralline algae on the slope; rubble, hard coral, and sand dominant substrata; dominant soft coral taxa was *Dendronepthya* spp. at 0–8 m and 9–20 m depths; hard coral cover = 13% at 7–9 m, 33% at 15–16 m; average hard coral cover

23%; light fishing pressure; important site for subsistence fishing, diving for commercially valuable resources, and trade. RCI = 215.46

51. Panasia Island, Louisiade Archipelago
Time: 0708 hours, dive duration 43 minutes; depth range 1–25 m; visibility approximately 10 m; temperature 26°C; no current; *site description*: sheltered fringing reef with gentle slope to deep water and with sandy substrate and coral bommies on the slope; sand and hard coral dominant substrata; hard coral cover = 35% at 7–8 m, 26% at 14–16 m; average hard coral cover 30.5%; light fishing pressure; moderate siltation; important site for subsistence fishing, diving for commercially valuable resources and trade. RCI = 184.45.

52. Pana Rai Rai Island, Louisiade Archipelago
Time: 1051 hours, dive duration 75 minutes; depth range 2–40 m; visibility approximately 15 m; temperature 26°C; moderate current; *site description*: exposed barrier type reef with vertical slope to deep water; hard coral and soft coral dominant substrata; hard coral cover = 55% at 6–8 m, 27% at 15–16 m; average hard coral cover 41%; light fishing pressure; important site for subsistence fishing and diving for commercially valuable resources. RCI = 201.68.

53. Jomard Entrance, Pana Waipona Island, Louisiade Archipelago
Time: 1300 hours, dive duration 47 minutes; depth range 1–42 m; visibility approximately 20 m; temperature 26°C; no current; *site description*: exposed fringing reef with very steep slope to deep water with caves and gullies leading to reef flat; soft coral and hard coral dominant substrata; hard coral cover = 48% at 7–8 m, 34% at 16–18 m; average hard coral cover 41%; light fishing pressure; important site for subsistence fishing and diving for commercially valuable resources. RCI = 222.53.

54. Punawan Island, Bramble Haven, Louisiade Archipelago
Time: 1556 hours, dive duration 48 minutes; depth range 2–25 m; visibility approximately 10–15 m; temperature 26°C; no current; *site description*: moderately exposed fringing/lagoon type reef with gentle slope to deep water with sand and coral bommies in the shallows and coral ridges running horizontally across the slope; hard coral and sand dominant substrata; hard coral cover = 52% at 7–8 m, 45% at 13–15 m; average hard coral cover 48.5%; light fishing pressure; important site for beche-de-mer and *Trochus* spp. harvesting; also temporary settlement for fishermen from Brooker Island for beche-de-mer processing. RCI = 244.48.

55. Bently Island, southern Engineer Group
Time: 1121 hours, dive duration 50 minutes; depth range 5–35 m; visibility approximately 15 m; temperature 26–27°C; no current; *site description*: moderately exposed fringing reef, sloping at angle of about 50° between depths

of 5–40 m; hard coral and rubble dominant substrata; hard coral cover = 55% at 7–8 m, 32% at 14–16 m; average hard coral cover 43.5%; light fishing pressure; important site for subsistence and commercial fishing. RCI = 219.98.

56. North Point, Basilaki Island
Time: 1453 hours, dive duration 56 minutes; depth range 1–35 m; visibility approximately 12–15 m; temperature 27°C; no current; slight turbidity; *site description*: sheltered fringing reef with moderate slope to deep water; hard coral and sand dominant substrata; hard coral cover = 70% at 7–9 m, 30% at 16–17 m; average hard coral cover 50%; light siltation and fishing pressure; important site for subsistence fishing. RCI = 225.79.

57. Negro Head, Sideia Island
Time: 1053 hours, dive duration 61 minutes; depth range 1–31 m; visibility approximately 10 m; temperature 26°C; no current; *site description*: sheltered fringing reef with very steep slope to deep water with caves and sand/rubble patches on slope; hard coral dominant substrata; hard coral cover = 67% at 7–8 m, 48% at 14–15 m; average hard coral cover 57.5%; light siltation and fishing pressure; important site for subsistence fishing; green turtle and grey reef shark seen. RCI = 253.31.

RESULTS

Reef condition
Data used for determining Reef Condition Index is presented in Appendix 3. The hypothetical maximum RCI, as explained previously, is 300. During the current survey, values ranged between 143.49 and 258.82. The top 10 sites for reef condition are presented in Table 2.2. These are sites that have the best combination of coral and fish diversity, as well as being relatively free of damage and disease.

Table 2.3 provides a frequency distribution of relative condition categories (see Methods section above for explanation).

Coral cover
Data for coral cover and other main substrate types are presented in Appendix 4. Percentage cover of live hard corals ranged from 13–85%, generally with an average between 30–50%. The highest coral cover was recorded at Urasi Island, Amphlett Group (85% for shallow transect) and Swinger Opening, Rossel Island (76% for shallow transect). The richest coral cover was generally recorded on the shallow (8 meter depth) transects.

Coral bleaching
Coral bleaching was present at every site (1–28) in Collingwood and Goodenough bays, the D'Entrecasteaux Islands, and Amphlett Islands. The most severe bleaching was at Jackdaw Channel near East Cape (site 1). Damage at most other sites was not serious. In most cases it appeared to

Table 2.2. Top 10 sites for general reef condition.

Site No.	Location	Fish species	Coral species	Condition points	RCI
29	Gabugabutau Island, Conflict Group	235	121	180	258.82
57	Negro Head, Sideia Island	209	120	190	253.31
39	Rossel Passage	224	106	200	252.45
54	Puawan Island, Bramble Haven	225	107	180	244.48
43	Swinger Opening, Rossel Island	225	106	180	243.71
4	Sanaroa Island, D'Entrecasteaux Islands	260	77	180	233.91
12	Kwatota Island, Amphlett Island	216	91	190	233.51
37	Marx Reef, Tagula Island	205	93	190	231.12
56	North Point, Basilaki Island	203	87	190	225.79
53	Jomard Entrance, Louisiade Archipelago	198	91	180	222.53

Table 2.3. Distribution of relative condition categories based on RCI values.

Relative condition	No. sites	Percentage of sites
Extraordinary	6	10.52
Excellent	12	21.05
Good	10	17.54
Moderate	21	36.84
Poor	8	14.03
Very Poor	0	0.00

be very recent, having occurred within days, or at most 2–3 weeks, of our visit (see Davis, *et al.,* 1997). The worst affected areas included the vicinity of Noapoi Island, Amphlett Group (site 10), Mukawa Bay, Cape Vogel area (site 14), and reefs in the southern portion of Collingwood Bay (sites 15–17). Bleaching was rarely observed during the remainder of the survey (sites 29–57) and negligible in extent. There was a definite correlation between warm sea temperatures and the occurrence of bleaching. The average water temperature in the areas affected by bleaching was 28.4° C compared to 27.0° C for non-bleached areas.

Coral pathogens and predators
The coral-feeding mollusc *Drupella* was noted at most sites, but damage was not serious. Coral disease was also prevalent at many sites, with the worst affected area being the Amphlett Islands. Crown-of-thorns starfish, another coral-feeding species, was extremely rare. Less than five animals were observed during the entire survey.

Cyclone damage
Wave damage was periodically noted, but was most severe at two sites in the Louisiades: Tawal Reef near Cormorant Channel (site 30) and Wanim Island, Calvados Chain (site 36).

Siltation
Coastal fringing reefs were invariably affected by well above average levels of silt, the result of terrestrial runoff. The worst-affected areas were the southern edge of Goodenough Bay (sites 24–28), around the islands of the Calvados Chain (sites 31–34, 36, 51), Tagula Island (sites 38, 46), and Rossel Island (sites 40–42). The fact that our survey was conducted during the rainy season no doubt resulted in higher than usual amounts of siltation and lower than average underwater visibility. Additionally, the survey coincided with the beginning of gardening activity where the villagers are cutting and burning new gardens. This may have potentially contributed to sedimentation at some of the study sites. Visibility ranged between 3–25 m, with an average value of about 10 m. The best visibility and least siltation was generally encountered at outer reef and passage sites.

Eutrophication
Extensive damage was observed on a patch reef in Rossel Lagoon at the mouth of Yonga Bay (site 41). The reef was almost entirely smothered by various algae, including large amounts of *Padina*. Extensive algae on reefs can be indicative of nutrient input or lack of herbivores. Judging by the fish data, it is plausible the cause may be eutrophication.

DISCUSSION

Reef condition
In general Milne Bay's reefs are in very good shape, particularly when compared to other parts of the Coral Triangle. Figure 1 compares the mean RCI for the current survey with

that of the Togean-Banggai Islands, off eastern Sulawesi, Indonesia (1998 RAP). An unpaired t-test revealed the mean for Milne Bay (199.32 ± 3.76) to be significantly greater than that recorded for the Togean-Banggai Islands (179.87 ± 4.02) at a 5% level of significance (Prob. [2-tail]: 0.0006, df: 102).

The superior reef condition of Milne Bay compared to previously surveyed areas in Indonesia and the Philippines probably is a result of its lower population density. Moreover, destructive fishing methods, notoriously rampant in Indonesia and the Philippines, have been rarely used in Milne Bay Province. Dynamite fishing in particular has severely impacted the reef environment in these countries. Fortunately, this type of fishing is not common, although there are periodic reports of its use, mainly in other parts of PNG (mostly in the Port Moresby area). The use of cyanide is another destructive fishing method that is commonly used in Indonesia and the Philippines, especially to catch large groupers and Napoleon Wrasse. This method was briefly introduced to Milne Bay Province about four years ago, but

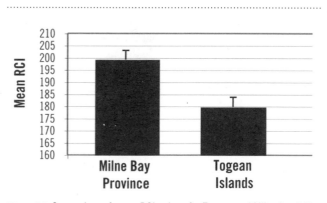

Figure 2.1. Comparison of mean RCI values for Togean and Milne Bay RAP surveys.

Table 2.4. Mean RCI values recorded for major areas within Milne Bay Province (2000 survey).

Major area	No. sites	Mean RCI
Conflict Group-Louisiade Archipelago	26	206.58
Cape Vogel area	6	203.07
Amphlett Islands	7	187.98
Goodenough Bay and East Cape region	7	169.82
Basilaki-Sideia Islands	2	239.55
South Collingwood Bay	3	187.07
D'Entrecasteaux Islands	5	194.68

the government summarily shut down the foreign company involved when the use of cyanide was discovered.

Table 2.4 presents a comparison of mean RCI values for the major geographical areas within Milne Bay Province as determined by the 2000 survey. The highest mean value was recorded for the two sites on northern Basilaki and Sideia Islands. Additional survey effort in this general area would be desirable to confirm its apparent richness. It is certainly not possible to draw conclusions on the basis of only two sites.

Otherwise, the most impressive area for reef condition was the combined Louisiade Archipelago-Conflict Group. However, this result is influenced by the preponderance of outer reef sites, which characteristically show high RCI values (Table 2.5).

Bleaching

Coral bleaching is a term used for the sudden loss of symbiotic zooxanthellae harbored by live soft and hard corals. It aptly describes the first symptom—the loss of symbiotic algae (zooanthellae) that live within the tissue of corals and other marine invertebrates. Not only are these single-celled plants responsible for the coral's normal color—usually shades of green, yellow, gray, and brown—but they also provide a valuable source of nutrients—primarily sugars that are "leaked" directly into the tissue of the host corals.

Scientists have been aware of the bleaching phenomenon for the past two decades and have now accumulated a wealth of data. In the 1980s it was first noted that global episodes of bleaching were correlated with strong El Niño events—pronounced shifting of oceanic water masses that directly affect our weather patterns. One of the effects of the strong El Niño is a 1–2° increase in tropical water temperatures. Researchers now believe that elevated temperature during the El Niño events is the primary factor responsible for triggering coral bleaching (Hoegh-Guldberg, 1999).

Research by Dr. Ove Hoegh-Guildberg and the staff of the Coral Reef Research Institute at the University of Sydney, Australia, reveals that tropical sea temperatures have increased by almost 1°C over the past 100 years and are now increasing at the alarming rate of 1–2° per century. Reef-building corals are now living perilously close to their upper limit of tolerance to potentially lethal temperatures. When the upper limit is exceeded, which often happens as a result of strong El Niño, the symbiotic algae become vulnerable to

Table 2.5. Mean RCI values for major habitats in Milne Bay Province.

Major habitat type	No. sites	Mean RCI
Outer reefs and passages	16	222.15
Isolated platform reefs	6	196.04
Fringing reefs	24	194.90
Lagoon reefs and patches	10	183.56

damage by light, which cause them to be expelled from the coral tissue, resulting in bleaching.

Increased sea temperatures have only had adverse affects since about 1980, when El Niño-boosted temperatures began to exceed the thermal tolerance of corals and their symbiotic algae. Over the past 20 years this critical upper limit (about 29–30°C at most locations) has been exceeded on numerous occasions, culminating in the devastating mass bleaching event of 1998. Projections for the future based on highly accurate models of global climate dramatically depict a scenario of wholesale mortality of the world's coral reefs. Bleaching events are projected to increase in frequency until they become yearly events by the year 2050. Projections indicate that lethal temperatures will be regularly exceeded by normal seasonal changes in water temperature in just 20–40 years from now.

So what can be done to stem this disturbing trend? This is a complex question that has no simple answer, but one obvious measure is to stem the tide of greenhouse gas emissions. Industrial and motor vehicle pollution and uncontrolled burning of forests are among the major factors contributing to global warming. Somehow, a cooperative program, involving all nations, rich and poor, must be undertaken to curb this trend before it's too late. Otherwise, we may be witnessing the beginning of the end of coral reefs on the grand scale they now occupy.

Milne Bay reefs appear to have been variably effected by recent bleaching activity (see Davis, et al., 1997). During the current survey it was consistently observed in northern areas (sites 1–28), and conspicuously absent from southern sites (29–30). Fortunately, in nearly every case where bleaching was noted, the damage was light, often involving sections of larger coral formations. The bleaching was usually recent, and there was good possibility of recovery. The only exception was the Jackdaw Channel area (site 1) where huge sections of the reef were dead, apparently the result of a bleaching episode within the past year.

As stated above, bleaching is a major threat to every coral reef area. The correlation between temperature and bleaching would indicate that northern sections of Milne Bay Province are more vulnerable than southern sections, where sea temperatures are cooler. For example, the average temperature for the combined 1997–2000 surveys for sites north of latitude 10°30' North was 28.27°C (N = 63) compared to 26.70 (n = 47) for sites south of this latitude. Unusually cool temperatures between 22–25°C were recorded during 1997 at sites in the vicinity of Samarai Island and along the southern shores of Sideia and Basilaki Islands. Cool temperatures in this region were possibly the result of upwelling.

Terrestrial runoff and siltation

Runoff and consequent siltation from adjacent areas subjected to deforestation could also pose a threat to Milne Bay's coral reef environment. Analysis of the magnitude of this threat is beyond the scope of the present Marine RAP, but the implication of unregulated logging on the province's coral reefs needs to be considered when planning or implementing a reef conservation and management program.

CONCLUSIONS

Milne Bay's reefs are in remarkably good condition, especially compared to other areas in the Coral Triangle, most notably Indonesia and the Philippines. But there is a danger of becoming complacent and failing to ensure that reefs are properly conserved for the use and enjoyment of future generations. It is vitally important to install effective management plans and conservation guidelines now, while reefs are still healthy and thriving. One only has to look immediately westward of PNG to appreciate the dire consequences of failure to properly manage coral reef resources.

REFERENCE

Davies, J. M., R. P. Dunne, and B. E. Brown. 1997. Coral bleaching and elevated seawater temperature in Milne Bay Province, Papua New Guinea, 1996. Marine and Freshwater Research 48(6): 513–516.

Hoegh-Guldberg, O. 1999. Climate change, coral bleaching, and the future of the world's coral reefs. Marine and Freshwater Research 50: 839–866.

Chapter 3

Molluscs of Milne Bay Province, Papua New Guinea

Fred E. Wells and Jeff P. Kinch

SUMMARY

- This report presents information on the molluscs collected at 28 sites surveyed in Milne Bay Province, Papua New Guinea, from 29 May to 10 June 2000. As many habitats as possible were examined at each site to develop as comprehensive a species list of the molluscs present in the limited time available.

- A total of 643 species of molluscs were collected: 482 gastropods, 155 bivalves, 2 scaphopods, 2 cephalopods, and 2 chitons.

- Diversity was high, and consistent with molluscan diversity recorded on other surveys in the coral triangle, which were undertaken for similar numbers of collecting days.

- Combined with the results of a similar Marine RAP survey undertaken in October–November 1997, 945 species of molluscs have been recorded in Milne Bay Province.

- The number of species collected per site ranged from 34 to 119, with a mean of 74.0 ± 4.4. Higher diversity was recorded at sites with more variable habitat types.

- A number of species were widespread, occurring at 15 or more sites. Several (*Coralliophila neritoidea, Drupella cornus, D. ochrostoma, Pyrene turturina, Tridacna squamosa, Turbo petholatus* and *Tectus pyramis*) live on or in close association with the coral, and others (*Pedum spondyloidaeum, Lithophaga* sp., *Arca avellana,* and *Tridacna crocea*) actually burrow into the coral. *Rhinoclavis asper* lives in sandy areas between the corals. Most species (78%) occurred at five or fewer sites.

- The most abundant species at each site were generally burrowing arcid bivalves, *Pedum spondyloidaeum, Lithophaga* sp., and *Coralliophila neritoidea. Tridacna crocea* was abundant at several of the sites.

- The 16 sites along the mainland coastline of Papua New Guinea had a mean of 66.0 ± 5.1 species per site. Diversity was higher (83.0 ± 14.6) at the four sites in the D'Entrecasteaux Islands. The highest diversity was 85.3 ± 7.9 in the Amphlett Islands. With the considerable range in number of species collected at each site, none of the three areas differed significantly from the mean of 74.0 ± 4.4 for all sites examined on the survey.

- Distribution patterns were determined for 258 species (belonging to well-documented groups): 207 species are widespread in the Indo-West Pacific, and 51 species are widespread in the Western Pacific. None of the species whose range was determined is endemic to Papua New Guinea.

- The history of the giant clam fishery in Milne Bay Province is briefly discussed. The fishery is closed at present. Recommendations made by Munro (1989) and Kinch (in press a and b) should be closely examined before any proposals are made to reopen the fishery.

INTRODUCTION

In October–November 1997 Conservation International (CI) conducted a Marine Rapid Assessment (RAP) survey of coral reefs in Milne Bay Province, Papua New Guinea. The goal of the expedition was to develop information on the biodiversity of three key animal groups—corals, fish, and molluscs—for use in assessing the importance of the reefs for conservation purposes. Goals, methodology, and results of the expedition are described in Allen et al. (1998); Wells (1998) described the molluscs. Following the success of the initial survey, additional surveys were conducted in the Calamianes Islands, Philippines, and the Togean and Banggai Islands, Indonesia. Molluscs were described by Wells (2000 and 2001).

The original expedition to Milne Bay Province confirmed the view of CI that coral reefs in Milne Bay Province of PNG have a high species diversity of the target groups selected for survey. In addition, they are largely undisturbed by adverse direct human activities such as overfishing, cyanide fishing, dynamiting and also indirect practices such as land clearing, which increases turbidity, pollution, use of antifoulants, etc. Plans for conserving Milne Bay Province's marine resources and environments are proceeding rapidly, and the Marine RAP team was asked to return to Milne Bay Province in May 2000 to examine areas not visited on the first survey.

In addition to their importance for conservation purposes, the Marine RAP surveys provide an increasing dataset on biodiversity of the three target groups on reefs in the Indo-West Pacific. This complements work done in a variety of areas of the eastern Indian Ocean by the Western Australian Museum.

METHODS

Molluscs were assessed during the survey from 29 May to 10 June 2000, with a total of 28 sites being examined. All sites, except site 13, were surveyed by scuba diving. Each site was examined by starting at depths of 20–40 m and working up the reef slope. Most of the time was spent in shallow (< 6 m) water, as the greatest diversity of molluscs occurs in this region, and the shallow depth maximizes diving time. To obtain as many species as possible, all habitats encountered at each site were examined for molluscs: living coral, the upper and lower surfaces of dead coral, shallow and deep sandy habitats, and intertidal habitats. For the same reason, no differentiation was made between species collected alive

or as dead shells, as the former occupants of dead shells would have been living at the site. Site 13 was an isolated coral atoll with an intertidal rock platform exposed by low tide. This site was examined by collecting in the exposed intertidal habitats and collecting beach drift.

This collecting approach allows the rapid assessment of the diversity of a wide variety of mollusc species. However, it is not complete. For example, no attempt was made to break open the corals to search for boring species, such as *Lithophaga*. Nor were micro molluscs sampled. However, as one person undertook the sampling of molluscs on the CI Marine RAPs, and many of those of the Western Australian Museum (WAM), there is a good indication of relative diversity of molluscs collected on the expeditions to various areas.

A variety of standard shell books and field guides were available for reference during the expedition. Most species were identified according to these texts, which included: Cernohorsky (1972); Springsteen and Leobrera (1986); Lamprell and Whitehead (1992); Brunckhorst (1993); Colin and Arneson (1995); Gosliner et al. (1996); Lamprell and Healy (1998); and Wells and Bryce (2000).

Specimens of small species were retained in plastic vials or bags and the tissue removed with bleach. These were taken to the WAM where they were identified using the reference collections of the Museum and specialist texts and papers on particular groups. Representatives of these species were deposited in the WAM. A set of reference materials of a number of the small species was also deposited in the National Museum of Papua New Guinea, Port Moresby.

RESULTS AND DISCUSSION

Despite the short time period available for mollusc sampling (a total of 11 collecting days), a diverse molluscan fauna was collected (Table 3.1). This consisted of a total of 643 species representing five molluscan classes: 482 gastropods, 155 bivalves, 2 scaphopods, 2 cephalopods, and 2 chitons (Table 3.1).

Table 3.1. Number of families, genera, and species of molluscs collected during the survey.

Class	Families	Genera	Species
Gastropoda	74	200	482
Bivalvia	30	92	155
Scaphopoda	2	2	2
Cephalopoda	2	2	2
Acanthopleura	2	2	2
Total	110	298	643

The survey compares favourably with the previous Marine RAP Surveys, where a range of 541 to 651 species was collected (Table 3.2). In particular, the present Milne Bay survey recorded 636 species, approximately 100 more than the survey of the Togean and Banggai Islands in Indonesia; both surveys were for 11 collecting days. The second Milne Bay survey collected approximately the same number of species as the 1997 RAP survey of Milne Bay and the survey of the Calamianes Islands, Philippines, both of which were for significantly longer periods of time. The present Milne Bay survey also compares favourably with similar collections that have been made in Western Australia and nearby areas by the WAM. Diversity recorded during the second Milne Bay survey was higher than all except one of the WAM surveys, the 655 species collected in the Muiron Islands and eastern Exmouth Gulf. The Muiron Islands expedition (12 days) was for a similar length of time as the 2000 Milne Bay survey, but had two mollusc collectors. In addition, the Muiron Island survey examined not only molluscs in the coral reefs of the Muiron Islands, but also the extensive shallow mudflats and mangrove communities of the eastern portion of Exmouth Gulf.

The second Milne Bay survey confirmed the high diversity of molluscs recorded on the previous 1997 CI survey. Combined, the two expeditions recorded a total of 945 species of marine molluscs. It should be emphasized that further collecting would undoubtedly reveal additional coral reef species. Additionally, habitats such as mangroves and mudflats have not yet been surveyed in Milne Bay Province.

Table 3.3 shows the total number of molluscs collected at each site varied from 34 to 119, with a mean of 74.0 ± 4.4 (S.E.). The fewest species were collected on vertical coral faces on isolated patch reefs (sites 2 and 9, with 48 and 55 species respectively) or on undersea cliffs (sites 14, 16, 24, and 25 with 58, 34, 41, and 53 species respectively). These sites lacked habitat diversity, and had few areas of high mollusc diversity. In particular, there was no sand available and few dead coral slabs to turn over and examine for molluscs. Site 23, at Mosquito Island, had the requisite habitat diversity. Unfortunately, due to the time of year that the survey was conducted, the sea was rough and the boat was unable to anchor. For this reason, the dive was restricted to one hour, and only 58 species were collected. Site 28 was visited late on the final collecting day of the expedition. Only an hour of

Table 3.2. Numbers of mollusc species collected during previous Marine RAP surveys undertaken by Conservation International and similar surveys by the Western Australian Museum.

Location	Collecting days	Mollusc species	Reference
Marine RAP Surveys			
Milne Bay, PNG	11	643	Present survey
Togian-Banggai Islands, Indonesia	11	541	Wells, in press a
Calamian Group, Philippines	16	651	Wells, 2001
Milne Bay, PNG	19	638	Wells, 1998
Western Australian Museum Surveys			
Cocos (Keeling) Islands	20	380 on survey; total known fauna of 610 species	Abbott, 1950; Maes, 1967; Wells, 1994
Christmas Island (Indian Ocean)	12 plus accumulated data	313 on survey; approx. 520 total	Iredale, 1917; Wells and Slack-Smith, 1988; Wells et al., 1990
Ashmore Reef	12	433	Wells, 1993; Willan, 1993
Cartier Island	7	381	Wells, 1993
Hibernia Reef	6	294	Willan, 1993
Scott/Seringapatam Reef	8	279	Wilson, 1985; Wells and Slack-Smith, 1986
Rowley Shoals	7	260	Wells and Slack-Smith, 1986
Montebello Islands	19	633	Preston, 1914; Wells et al., 2000
Muiron Islands and Exmouth Gulf	12	655	Slack-Smith and Bryce, 1995
Bernier and Dorre Islands, Shark Bay	12	425	Slack-Smith and Bryce, 1996
Abrolhos Islands	Accumulated data	492	Wells and Bryce, 1997
Other surveys			
Chagos Islands	Accumulated data	384	Shepherd, 1984

collecting could be done before darkness prevented further activity. Shallow and intertidal habitats, which were available at the site, could not be examined at all.

The sites with the greatest diversity of molluscs (Table 3.4) were those with the greatest habitat diversity. In particular these sites had shallow sand in addition to subtidal corals and intertidal rocks. Shallow sand is important both because of the species which live within it and because dead shells accumulate there from adjacent coral habitats.

Table 3.3. Total number of mollusc species collected at each site.

Site	No. species	Site	No. species	Site	No. species
1	90	11	99	21	84
2	48	12	94	22	74
3	100	13	71	23	58
4	113	14	58	24	41
5	71	15	64	25	53
6	70	16	34	26	59
7	112	17	50	27	119
8	67	18	81	28	58
9	55	19	69	Mean	74.0 ± 4.4
10	114	20	65		

Table 3.4. Ten richest sites for mollusc diversity.

Site	Location	No. species
27	Scrub Islet, off Sanaroa Island, D'Entrecasteaux Islands	119
10	Rock Islet near Noapoi Island, Amphlett Group	114
4	Kuvira Bay, South Goodenough Bay	113
7	Toiyana Island, Amphlett Group	112
3	Sebulgomwa Point, Fergusson Island, D'Entrecasteaux Islands	100
11	East side of Kwatota Island, Amphlett Group	99
12	Northwest corner of Kwatota Island, Amphlett Group	94
1	Cobb's Cliff, Jackdaw Channel	90
18	Ipoteto Island, Kibirisi Point, Cape Vogel	81
21	South of Ragrave Point, Cape Vogel	84

As indicated above, a total of 28 sites were examined during the survey. Most species (494 or 78%) were collected at five sites or less; very few species were widespread. In fact, only 14 of the 523 species were collected at 15 or more sites (Table 3.5). These species can be used to characterize the dominant species on the reef. Several species (*Coralliophila neritoidea, Drupella cornus, D. ochrostoma, Pyrene turturina, Tridacna squamosa, Turbo petholatus,* and *Tectus pyramis*) live on or in close association with the coral, and others (*Pedum spondyloidaeum, Lithophaga* sp., *Arca avellana,* and *Tridacna crocea*) actually burrow into the coral. *Rhinoclavis asper* lives in sandy areas between the corals.

The fact that these species were each found at 15 or more sites does not mean that they were all abundant, as many of the records are based on one or a few dead shells found at the site. The most abundant species at each site were generally burrowing arcid bivalves, *Pedum spondyloidaeum, Lithophaga* sp., and *Coralliophila neritoidea. Tridacna crocea* was abundant at several of the sites.

The sites were divided into three geographical areas (Table 3.6). Most sites (16) were along the mainland coastline of Papua New Guinea. These sites had a mean of 66.0 ± 5.1 species per site. Diversity was higher (83.0 ± 14.6) at the four sites in the D'Entrecasteaux Islands. The highest diversity was 85.3 ± 7.9 in the Amphlett Islands. With the considerable range in number of species collected at each site, none of the three areas differed significantly from the mean of 74.0 ± 4.4 for all sites examined during the survey.

The molluscs collected on the survey were separated into groups based on their biogeographic distributions (Table 3.7). As the ranges of many species are poorly known, only those groups that have been examined by recent revisions

Table 3.5. Most widespread species of molluscs.

Species	Class	No. sites
Coralliophila neritoidea	Gastropoda	25
Pyrene turturina	Gastropoda	24
Pedum spondyloidaeum	Bivalvia	23
Drupella cornus	Gastropoda	21
Rhinoclavis asper	Gastropoda	20
Tridacna squamosa	Bivalvia	19
Turbo petholatus	Gastropoda	18
Tectus pyramis	Gastropoda	18
Cypraea lynx	Bivalvia	18
Lithophaga sp.	Bivalvia	18
Arca avellana	Bivalvia	18
Tridacna crocea	Bivalvia	17
Drupella ochrostoma	Gastropoda	15
Conus miles	Gastropoda	15

Table 3.6. Geographical distribution of molluscs in the three regions covered by the survey of Milne Bay Province.

Geographic area	Sites	No. of species		
		Min.	Max.	Mean ± 1 S.E.
Mainland	1, 14–28	33	119	66.0 ± 5.1
D'Entrecasteuax Islands	2–5	48	113	83.0 ± 14.6
Amphlett Islands	6–13	48	101	85.3 ± 7.9
Overall	1–28	34	119	74.0 ± 4.4

Table 3.7. Geographical distribution of selected species of molluscs collected during Milne Bay Province survey.

Geographic area	No. species	Percentage
Indo-West Pacific	207	80
Western Pacific	51	20
Endemic to Papua New Guinea and the Coral Sea	0	0
Total	258	100

were included. The great majority of species studied (207 of 258) are widespread throughout the Indo-West Pacific. Fifty-one species are widespread in the Western Pacific Ocean. None of the species studied is endemic to the Papua New Guinea.

A number of commercially important mollusc species occurred widely at the sites surveyed. Five species of the spider shell genus *Lambis* were recorded, with *L. millepeda* occurring at 11 sites. Six species of giant clams (*Tridacna*) were recorded. The most widespread were *T. squamosa* (19 sites) and *T. crocea* (17 sites). However, most of these species were recorded on the basis of one or a few individuals per site, and in some cases the records were based on dead shells. There was a pile of approximately 12 *T. squamosa* and *Hippopus hippopus* in the shallows that had been obviously eaten by people. There were also isolated individuals of both species around the island.

Giant clams are a major fishery group in the Pacific Ocean. The large size of individual animals, their shallow water habitat, and their longevity means the species can be rapidly fished out in local areas. This has happened in many parts of the Pacific Ocean.

Commercial fisheries for giant clams developed in Milne Bay Province in the wake of the reduction of illegal fishing by Taiwanese vessels and in response to sustained demand. Poaching reached a prolonged peak from 1967 to 1981 (Kinch, 1999). It declined after 1981 because of depleted

stocks, strong international pressure, and increased surveillance. The Milne Bay Fisheries Authority (MBFA), established in 1979 (Munro, 1989), began export of giant clams from the province in 1983 (Lokani and Ada, 1998). A ban on the purchase and export of wild-caught giant clam meat was later placed in May 1988 but lifted in May 1995. During the ban, some regeneration of giant clam stocks occurred, which provided an incentive for a local fishing company to commence harvesting and exports. This ban was put back in place in 2000 when it was found that a local fishing company was infringing on its licensing arrangements (Kinch, 2001 and in press).

Chesher carried out the first stock abundances estimated for tridacnid species in 1980 in the southeast of Milne Bay Province. He stated that prior to commercial harvesting, this area contained an estimated overall density of 39/ha for all species of giant clams (Chesher, 1980). In 1996, a stock assessment by the South Pacific Commission (SPC) and the PNG National Fisheries Authority (NFA) was carried out in the Engineer and Conflict Groups of islands, an area further northeast than that surveyed by Chesher. Sixty-three sites were surveyed in this area. From extrapolation of these data it was suggested by Ledua et al. (1996) that approximately 98 percent of the stock of *Tridacna gigas* throughout Milne Bay Province had been wiped out since the opening of the Milne Bay Fishing Authority in the early 1980s, with overall stock density of all species estimated to be down by 82.35 percent of the original population. The overall density of all species was estimated to be only 0.5/ha (Ledua et al., 1996).

Given the importance of giant clams to the people of Milne Bay Province, and following the results of the 2000 RAP and previous stock assessment attempts, a more thorough and comprehensive assessment of *Tridacnid* stocks was required. The Australian-based Commonwealth Scientific and Industrial Organization (CSIRO), the National Fisheries Authority (NFA), and CI conducted a sedentary resources stock assessment, which included giant clams, during the months of October and November 2001, with 1126 sites surveyed throughout Milne Bay Province. *Tridacna gigas* abundances ranged from 0 to 1.32/ha across the Province with a mean density of 0.82/ha. Throughout the Province, mean densities for the other species were *Tridacna maxima* at 1.79/ha, *T. derasa* at 0.34/ha, *T. squamosa* at 1.37/ha, *T. crocea* at 14.85/ha, and *Hippopus hippopus* at 0.41/ha. The species harvested for commercial use had a mean density of 4.32/ha (Kinch, 2002). The results from the CSIRO/NFA/CI stock assessment indicated clearly that stock levels are very low and have been heavily depleted across Milne Bay Province.

From January to the end of September 1999, a local fishing company purchased 697 kg of giant clam muscles—mostly *Tridacna gigas* and *T. derasa*—from Brooker Islanders (Table 8). During this period, the local fishing company was purchasing two sizes of giant clam muscle based on weight. A kilogram of specimens, each weighing under 400 g, earned 6 kina (US$ 1.6), and a kilogram of specimens weighing more than 400 g fetched 10 kina (US$ 2.7). Total purchases from January to September were broken down into 551 kg (or 1970 clams) of specimens under 400 g earning 3306 kina (US$ 915), and 146 kg (or 170 clams) earning 1460 kina (US$ 404) (Kinch, 1999). Of this volume, almost a third of the *T. gigas* were not full-grown adults.

In order to ensure the sustainability of all marine resources including giant clams in Milne Bay Province, effective management strategies must be implemented. Several approaches to assist the recovery of overfished tridacnid populations have been proposed. These include the establishment of Marine Protected Areas (MPAs), concentrating the remaining adult clams so that their reproduction can be facilitated by their closer proximity, and seeding cultured giant clams of sufficient size or in sufficient numbers and releasing these into the field to produce adult populations (Kinch, 2001; in press). The imposition of a ban on further commercial fishing or strict harvesting quotas over a single short season, coupled with the size restrictions, could also offer prospects for management.

ACKNOWLEDGMENTS

We would like to warmly acknowledge the support of the other participants in the cruise for their help in collecting specimens, exchanging ideas, and providing an enjoyable time. In addition, we very much appreciate the assistance provided by Mrs. Glad Hansen and Mr. Hugh Morrison of the Western Australian Museum in identifying material brought back to the Museum.

REFERENCES

Abbott, R. T. 1950. Molluscan fauna of the Cocos-Keeling Islands. Bulletin of the Raffles Museum 22: 68–98.

Allen, G. R. and S. A. McKenna (eds.). 2002. A RAP Biodiversity Assessment of the Coral Reefs of the Togean and Banggai Islands, Sulawesi, Indonesia. RAP Bulletin of Biological Assessment. Washington, DC: Conservation International.

Brunckhorst, D. J. 1993. The Systematics and Phylogeny of Phyllidiid Nudibranchs (Doridoidea). Records of the Australian Museum, Supplement 16: 1–107.

Cernohorsky, W. O. 1972. Marine Shells of the Pacific. Volume 2. Sydney, Australia: Pacific Publications.

Chesher, R. 1980. Stock Assessment: Commercial Invertebrates of Milne Bay reefs. Unpublished report, Fisheries Division, Dept of Primary Industries (Papua New Guinea).

Colin, P. L. and C. Arneson. 1995. Tropical Pacific Invertebrates. Beverley Hills, California: The Coral Reef Research Foundation and Coral Reef Press.

Gosliner, T. M., D. W. Behrens, and G. C. Williams. 1996. Coral Reef Animals of the Indo-Pacific. Monterey, California: Sea Challengers.

Iredale, T. 1917. On some new species of marine molluscs from Christmas Island, Indian Ocean. Proceedings of the Malacological Society of London 12: 331–334.

Kinch, J. P. 1999. Economics and Environment in Island Melanesia: A General Overview of Resource use and Livelihoods on Brooker Island in the Calvados Chain of the Louisiade Archipelago, Milne Bay Province, Papua New Guinea. Unpublished report for Conservation International-Papua New Guinea by the Department of Anthropology, University of Queensland, Australia.

Kinch, J. 2001. Clam Harvesting, the Convention on the International Trade in Endangered Species (CITES) and Conservation in Milne Bay Province. SPC Fisheries Newsletter. 99: 24–36.

Kinch, J. 2002. Giant Clams: Their Status and Trade in Milne Bay Province, Papua New Guinea. Traffic Bulletin 19(2): 67–75.

Lamprell, K. and J. M. Healy. 1998. Bivalves of Australia. Volume 2. Leiden, The Netherlands: Backhuys Publishers.

Lamprell, K. and T. Whitehead. 1992. Bivalves of Australia. Volume 1. Bathurst, Australia: Crawford House Press.

Ledua, E., S. Matoto, R. Lokani, and L. Pomat. 1996. Giant clam resource assessment in Milne Bay Province. Unpublished report prepared by the South Pacific Commission and the National Fisheries Authority (Papua New Guinea).

Lucas, J. 1994. The biology, exploitation and mariculture of giant clams (Tridacnidae). Reviews in Fisheries Science 2(3): 181–223.

Maes, V. O. 1967. The littoral marine molluscs of Cocos-Keeling Islands (Indian Ocean). Proceedings of the Academy of Natural Science of Philadelphia 119: 93–217.

Munro, J. 1989. Development of a giant clam management strategy for Milne Bay Province. Unpublished report, Department of Fisheries and Marine Resources (Papua New Guinea).

Preston, H. B. 1914. Description of new species of land and marine shells from the Montebello Islands, Western Australia. Proceedings of the Malacological Society of London 11: 13–18.

Sheppard, A. L. S. 1984. The molluscan fauna of Chagos (Indian Ocean) and an analysis of its broad distribution patterns. Coral Reefs 3: 43–50.

Slack-Smith, S. M. and C. W. Bryce. 1995. Molluscs. In: Hutchins, J. B., S. M. Slack-Smith, L. M. Marsh, D. S. Jones, C. W. Bryce, M. A. Hewitt, and A. Hill (eds.) Marine biological survey of Bernier and Dorre Islands, Shark Bay. Unpublished report, Western Australian Museum and Department of Conservation and Land Management. Pp. 57–81.

Slack-Smith, S. M. and C. W. Bryce. 1996. Molluscs. In: Hutchins, J.B., S. M. Slack-Smith, C. W. Bryce, S. M. Morrison, and M. A. Hewitt. 1996. Marine biological survey of the Muiron Islands and the eastern shore of Exmouth Gulf, Western Australia. Unpublished report, Western Australian Museum and Department of Conservation and Land Management. Pp. 64–100.

Springsteen, F. J. and F. M. 1986. Shells of the Philippines. Manila: Carfel Seashell Museum.

Wells, F. E. 1993. Part IV. Molluscs. *In*: Berry, P.F. (ed.). Faunal Survey of Ashmore Reef, Western Australia. Records of the Western Australian Museum Supplement 44: 25–44.

Wells, F. E. 1994. Marine Molluscs of the Cocos (Keeling) Islands. Atoll Research Bulletin 410: 1–22.

Wells, F. E. 1998. Marine Molluscs of Milne Bay Province, Papua New Guinea. *In:* Werner, T. and G. R. Allen (eds.). A rapid biodiversity assessment of the coral reefs of Milne Bay Province, Papua New Guinea. RAP Working Papers Number 11. Washington, DC: Conservation International. Pp. 35–38.

Wells, F. E. 2000. Molluscs of the Calamianes Islands, Palawan Province, Philippines. *In*: Werner, T. B. and G. R. Allen (eds.). A Rapid Marine Biodiversity Assessment of the Calamianes Islands, Palawan Province, Philippines. Bulletin of the Rapid Assessment Program 17. Washington, DC: Conservation International.

Wells, F. E. 2001. Molluscs of the Gulf of Tomini, Sulawesi, Indonesia. *In*: Allen G. R., and S. A. McKenna (eds.). A Rapid Biodiversity Assessment of the Coral Reefs of the Togean and Banggai Islands, Sulawesi, Indonesia. RAP Bulletin of Biological Assessment. Washington, DC: Conservation International.

Wells, F. E. In press. Centres of biodiversity and endemism of shallow water marine molluscs in the tropical Indo-West Pacific. Proc. Ninth International Coral Reef Symposium, Bali.

Wells, F. E. and C. W. Bryce. 1997. A preliminary checklist of the marine macromolluscs of the Houtman Abrolhos Islands, Western Australia. *In*: Wells, F. E. (ed). Proceedings of the seventh international marine biological workshop: The marine flora and fauna of the Houtman Abrolhos Islands, Western Australia. Perth: Western Australian Museum. Pp. 362–384.

Wells, F. E. and C. W. Bryce. 2000. Seashells of Western Australia. Perth: Western Australian Museum.

Wells, F. E., C. W. Bryce, J. E. Clarke, and G. M. Hansen. 1990. Christmas Shells: The Marine Molluscs of Christmas Island (Indian Ocean). Christmas Island Natural History Association.

Wells, F. E. and S. M. Slack-Smith. 1986. Part IV. Molluscs. *In*: Berry, P.F. (ed.). Faunal survey of the Rowley Shoals and Scott Reef, Western Australia. Records of the Western Australian Museum Supplement 25: 41–58.

Wells, F. E. and S. M. Slack-Smith. 1988. Part V. Molluscs. *In*: Berry, P.F. (ed). Faunal survey of Christmas Island (Indian Ocean). Unpublished report, Australian National Parks and Wildlife Authority. Pp. 36–48.

Wells, F. E., S. M. Slack-Smith, and C. W. Bryce. 2000. Molluscs of the Montebello Islands. *In*: Berry, P.F. and F. E. Wells (eds.). Survey of the marine fauna of the Montebello Islands, Western Australia. Records of the Western Australian Museum Supplement 59. Pp. 29–46.

Willan, R. C. 1993. Molluscs. *In*: Russell, B.C. and J. R. Hanley. The marine biological resources and heritage values of Cartier and Hibernia Reefs, Timor Sea. Unpublished report for Northern Territory Museum, Darwin.

Wilson, B. R. 1985. Notes on a brief visit to Seringapatam Atoll, North West Shelf, Australia. Atoll Research Bulletin 292: 83–100.

Chapter 4

Reef Fishes of Milne Bay Province, Papua New Guinea

Gerald R. Allen

SUMMARY

- A list of fishes was compiled for 57 sites, mainly in the following sections of Milne Bay Province: Collingwood Bay, Goodenough Bay, Amphlett Islands, and Louisiade Archipelago. The survey involved 75 hours of scuba diving to a maximum depth of 51 m.

- Milne Bay Province has one of the world's richest reef and shore fish faunas, consisting of approximately 1,109 species of which 798 (72%) were observed or collected during the present survey. The total includes 67 species, not recorded during the previous (1997) Marine RAP.

- A formula for predicting the total reef fish fauna based on the number of species in six key indicator families indicates that at least 1,300 species can be expected to occur in Milne Bay Province.

- Gobies (Gobiidae), wrasses (Labridae), and damselfishes (Pomacentridae) are the dominant families in Milne Bay Province in terms of number of species (124, 108, and 100 respectively) and number of individuals observed.

- Species numbers at visually sampled sites ranged from 140 to 260, with an average of 195. The average for 110 sites surveyed during 1997 and 2000 is 192.

- Fish diversity of Milne Bay Province greatly surpasses that of other areas in the Coral Triangle previously surveyed as of June 2000 by CI. It dominates (19 of 21 sites) the list of top fish sites for combined RAP surveys in Indonesia (Togean-Banggai), Papua New Guinea, and the Philippines.

- Two hundred or more species per site is considered the benchmark for an excellent fish count. This figure was achieved at 42% of Milne Bay sites compared to 19% of sites in the Togean-Banggai Islands, Indonesia, and 10.5% at the Calamianes Islands, Philippines.

- The Nuakata region and Conflict Group were the richest areas for reef fishes. Their dominance reflects a high percentage of outer-reef dropoffs, which consistently harbor a high number of fishes.

- Outer reefs and passages contain the highest fish diversity, with an average of 213 species per site. Other major habitats include fringing reefs (192 per site), isolated platform reefs (191 per site), and lagoon reefs (178 per site).

- The following species are known thus far only from reefs of Milne Bay Province: *Chrysiptera cymatilis* (Pomacentridae), *Cirrhilabrus pylei* (Labridae), *Novaculichthys* n. sp. (Labridae), *Ecsenius taeniatus* (Blenniidae), and *Trichonotus halstead* (Tichonotidae). However, future collecting in adjacent provinces will probably expand their ranges.

• Areas with the highest concentration of fish diversity and consequent high conservation and management potential include: the Nuakata area (1997 survey), Conflict Group, Cape Vogel area, and Bramble Haven.

INTRODUCTION

The principal aim of the fish survey was to provide a comprehensive inventory of the reef species inhabiting Milne Bay Province and their associated habitats. This segment of the fauna includes fishes living on or near coral reefs down to the limit of safe sport diving or approximately 50 m depth. It therefore excludes deepwater fishes, offshore pelagic species such as flyingfishes, tunas, and billfishes, and purely estuarine species.

The results of this survey facilitate a comparison of the faunal richness of Milne Bay Province with other parts of Papua New Guinea and adjoining regions. However, the list of fishes presented below is still incomplete, due to the rapid nature of the survey and the cryptic nature of many small reef species. Nevertheless, a basic knowledge of the cryptic component of the fauna in other areas and an extrapolation method that utilizes key "index" families can be used to predict Milne Bay's overall species total.

METHODS

The fish portion of this survey involved 75 hours of scuba diving by G. Allen to a maximum depth of 51 m. A list of fishes was compiled for 57 sites. The basic method consisted of underwater observations made, in most cases, during a single, 60–90 minute dive at each site. The name of each species seen was written on a plastic sheet attached to a clipboard. The technique usually involved a rapid descent to 10–51 m, then a slow, meandering path was traversed on the ascent back to the shallows. The majority of time was spent in the 2–12 m depth zone, which consistently harbors the largest number of species. Each visual transect included a representative sample of all available bottom types and habitat situations, for example rocky intertidal, reef flat, steep drop-offs, caves (utilizing a flashlight if necessary), rubble, and sand patches.

Only the names of fishes for which identification was absolutely certain were recorded. However, very few (less than about two percent) could not be identified to species. This high level of recognition is based on more than 25 years of underwater fish observation experience in the Indo-Pacific and an intimate knowledge of the reef fishes of this vast region as a result of laboratory and field studies.

The visual survey was supplemented with eight small collections procured with the use of the ichthyocide rotenone and several specimens selectively collected with a rubber-sling propelled multi-prong spear. The purpose of the rotenone collections was to flush out small crevice and subsand-dwelling fishes (for example eels and tiny gobies) that are rarely recorded with visual techniques.

The present report also includes a comprehensive list of reef fishes of Milne Bay Province based on the 1997 and 2000 CI surveys. It was supplemented by records received from Bob Halstead, former owner and operator of the dive-charter boat *Telita,* and Dr. John Randall, former curator of the Bishop Museum, Honolulu. Additionally, Rudie Kuiter, a highly competent Australian colleague who has authored several major fish books dealing with the Australasian region, contributed a number of records.

RESULTS

The total reef fish fauna of Milne Bay Province reported herein consists of 1,109 species. A total of 798 species belonging to 262 genera and 69 families were recorded during the present survey. In combination with the species reported by Allen (1998), the Milne Bay totals are now boosted to 1,109 species, 357 genera, and 93 families (Appendix 4). Most of the species were illustrated by Allen (1991 and 1993), Myers (1989), Kuiter (1992), and Randall et al. (1990).

General faunal composition

The fish fauna of Milne Bay Province consists mainly of species associated with coral reefs. The most abundant families in terms of number of species are gobies (Gobiidae), wrasses (Labridae), damselfishes (Pomacentridae), cardinalfishes (Apogonidae), groupers (Serranidae), butterflyfishes (Chaetodontidae), blennies (Blenniidae), surgeonfishes (Acanthuridae), snappers (Lutjanidae), and parrotfishes (Scaridae). These 10 families collectively account for 58.7 percent of the total reef fauna (Fig. 4.1).

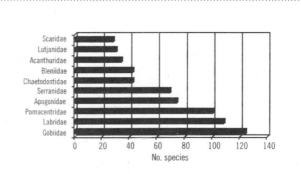

Figure 4.1. Ten largest families of Milne Bay.

The relative abundance of Milne Bay Province fish families is similar to other reef areas in the Indo-Pacific, although the ranking of individual families is variable as shown in Table 4.1. Even though the Gobiidae was the leading family, it was not adequately collected, due to the small size and cryptic habits of many species. Similarly, the moray eel family Muraenidae is consistently among the most speciose groups at other localities, and is no doubt abundant in Milne Bay Province. However, they are best sampled with rotenone due to their cryptic habits.

Habitats and fish diversity

Milne Bay Province has a rich marine ichthyofauna in comparison with other coral reef areas in the Indo-Pacific region. It is mainly composed of widely distributed elements that are recruited as postlarvae, after a variable pelagic stage. The total species present at a particular locality is ultimately dependent on the availability of food and shelter and the diversity of substrata. Coral and rocky reefs exposed to periodic strong currents are by far the richest habitat in terms of fish diversity. These areas provide an abundance of shelter for fishes of all sizes, and the currents are vital for supporting numerous planktivores, the smallest of which provide food for larger predators. Silty bays (often relatively rich for corals), mangroves, seagrass beds, and pure sand-rubble areas are consistently the poorest areas for fish diversity. The highest number of fish species were usually recorded at sites incorporating the following features: (1) predominantly coral or rock reef substratum, (2) relatively clear water, (3) periodic strong currents, and (4) presence of additional habitats (sand-rubble, seagrass, mangroves, etc.) in close proximity (i.e., within easy swimming distance of the primary coral reef habitat). The number of species found at each site is

indicated in Table 4.2. Totals ranged from 140 to 260, with an average of 195 species per site.

Table 4.2. Number of species observed at each site.

Site	Species	Site	Species	Site	Species
1	160	20	197	39	224
2	171	21	185	40	162
3	179	22	218	41	141
4	260	23	140	42	198
5	194	24	168	43	225
6	224	25	162	44	230
7	195	26	216	45	238
8	189	27	177	46	172
9	194	28	170	47	167
10	211	29	235	48	210
11	188	30	232	49	211
12	216	31	201	50	206
13	182	32	190	51	192
14	245	33	169	52	187
15	194	34	154	53	198
16	153	35	224	54	225
17	177	36	173	55	165
18	186	37	205	56	203
19	208	38	191	57	209

Table 4.1. Family ranking in terms of number of species for various localities in the Indo-Pacific region. Data for Kimbe Bay, New Britain, is from Allen and Munday (1994, unpublished), for Flores from Allen and Kuiter (1994, unpublished), for Komodo from Allen (1995, unpublished), for the Chagos Archipelago from Winterbottom et al. (1989), and for the Marshall Islands from Randall and Randall (1987).

Family	Milne Bay Province	Kimbe Bay Papua New Guinea	Flores, Indonesia	Komodo, Indonesia	Chagos Archipelago	Marshall Islands
Gobiidae	1st	2nd	1st	3rd	1st	1st
Labridae	2nd	2nd	2nd	2nd	2nd	2nd
Pomacentridae	3rd	1st	3rd	1st	3rd	4th
Apogonidae	4th	5th	4th	4th	6th	8th
Serranidae	5th	4th	5th	5th	4th	3rd
Chaetodontidae	6th	6th	7th	6th	11th	8th
Blenniidae	6th	8th	6th	8th	9th	6th
Acanthuridae	8th	7th	8th	7th	8th	7th
Lutjanidae	9th	9th	9th	9th	7th	18th
Scaridae	10th	10th	10th	10th	12th	10th

Richest sites for fishes

The 11 most speciose sites for fishes are indicated in Table 4.3. The average total for all sites (195) was remarkably high compared to other previously surveyed areas in the Coral Triangle (see discussion), lending support to the claim that Milne Bay Province is among the best areas for reef fishes in the entire Indo-Pacific region.

The statistics are even more impressive if the results of the 1997 and 2000 surveys are combined (Table 4.4).

Table 4.5 presents a reef fish fauna comparison of major geographical areas surveyed within Milne Bay Province during both the 1997 and 2000 surveys. The highest average number of species (220) was recorded for the Conflict Group and the area centered on Nuakata Island (including Hardman Island). The lowest value was recorded for the D'Entrecasteaux Islands.

Coral Fish Diversity Index (CFDI)

Allen (1998) devised a convenient method for assessing and comparing overall reef fish diversity. The technique essentially involves an inventory of six key families: Chaetodontidae, Pomacanthidae, Pomacentridae, Labridae, Scaridae, and Acanthuridae. The number of species in these families is totalled to obtain the Coral Fish Diversity Index (CFDI) for a single dive site, relatively restricted geographic areas (e.g., Milne Bay Province) or countries and large regions (e.g., Papua New Guinea or Great Barrier Reef).

CFDI values can be used to make a reasonably accurate estimate of the total coral reef fish fauna of a particular locality by means of regression formulas. The latter were obtained after analysis of 35 Indo-Pacific locations for which reliable, comprehensive species lists exist. The data were first divided into two groups: those from relatively restricted localities (surrounding seas encompassing less than 2,000 km^2) and those from much larger areas (surrounding seas encompass-

Table 4.4. Richest sites based on 1997 and 2000 surveys. Asterisk (*) indicates 1997 survey.

Site	Location	Total fish spp.
20*	Boirama Island	270
12*	East Irai Island	268
48*	Wahoo Reef	245
4	Scub Islet, Sanaroa Island	260
14	Mukawa, Cape Vogel	245
50*	Boia-boia Waga Island	243
13*	West Irai Island	241
45	West Barrier Reef, Rossel Island	238
21*	Nuakata Island	237
46*	Kathy's Corner, East Cape	236
29	Gabugabutau Island, Conflict Group	235
19*	Hardman Island	234
30	Tawal Reef, Louisiade Archipelago	232
44	Northwest Barrier Reef, Rossel Island	230
43	Swinger Opening, Rossel Island	225
54	Punawan Island, Bramble Haven	225
6	Catsarse Reef, Amphlett Islands	224
35	Sabari Island, Louisiade Archipelago	224
39	Rossel Passage	224

Table 4.3. Ten richest fish sites during the 2000 survey.

Site	Location	Total fish spp.
4	Scub Islet, Sanaroa Island	260
14	Mukawa, Cape Vogel	245
45	West Barrier Reef, Rossel	238
29	Gabugabutau Island, Conflict Group	235
30	Tawal Reef, Louisiade Archipelago	232
44	Northwest Barrier Reef, Rossel Island	230
43	Swinger Opening, Rossel Island	225
54	Punawan Island, Bramble Haven	225
6	Catsarse Reef, Amphlett Islands.	224
35	Sabari Island, Louisiade Archipelago	224
39	Rossel Passage	224

Table 4.5. Average number of fish species per site recorded for major areas within Milne Bay Province during the 1997 and 2000 surveys.

Area	Year	No. sites	Average no. species
Conflict Group	1997/2000	8	222
Nuakata Region	1997	8	220
Cape Vogel area	2000	6	207
Amphlett Islands	2000	7	202
Laseinie-Engineer Group	1997/2000	6	198
Louisiade Archipelago	2000	25	197
Goodenough Bay and East Cape region	1997/2000	14	193
China Strait to Basilaki Island	1997/2000	9	175
South Collingwood Bay	2000	3	175
D'Entrecasteaux Islands	1997/2000	22	173

ing more than 50,000 km^2). Simple regression analysis revealed a highly significant difference (P = 0.0001) between these two groups. Therefore, the data were separated and subjected to additional analysis. The Macintosh program Statview was used to perform simple linear regression analyses on each data set in order to determine a predictor formula, using CFDI as the predictor variable (x) for estimating the independent variable (y) or total coral reef fish fauna. The resultant formulae were obtained: (1). total fauna

of areas with surrounding seas encompassing more than 50,000 km^2 = 4.234(CFDI) - 114.446 (d.f = 15; R^2 = 0.964; P = 0.0001); (2). total fauna of areas with surrounding seas encompassing less than 2,000 km^2 = 3.39 (CFDI) - 20.595 (d.f = 18; R^2 = 0.96; P = 0.0001).

The CFDI regression formula is particularly useful for large regions, such as Indonesia and the Philippines, where reliable totals are lacking. Moreover, the CFDI predictor value can be used to gauge the thoroughness of a particu-

Table 4.6. Coral fish diversity index (CFDI) values for restricted localities, number of coral reef fish species as determined by surveys to date, and estimated numbers using the CFDI regression formula (refer to text for details).

Locality	CFDI	No. reef fishes	Estimated reef fishes
Milne Bay, Papua New Guinea	337	1109	1313
Maumere Bay, Flores, Indonesia	333	1111	1107
Togean and Banggai Islands, Indonesia	308	819	1023
Komodo Islands, Indonesia	280	722	928
Madang, Papua New Guinea	257	787	850
Kimbe Bay, Papua New Guinea	254	687	840
Manado, Sulawesi, Indonesia	249	624	823
Capricorn Group, Great Barrier Reef	232	803	765
Ashmore/Cartier Reefs, Timor Sea	225	669	742
Kashiwa-Jima Island, Japan	224	768	738
Scott/Seringapatam Reefs, Western Australia	220	593	725
Samoa Islands, Polynesia	211	852	694
Chesterfield Islands, Coral Sea	210	699	691
Sangalakki Island, Kalimantan, Indonesia	201	461	660
Bodgaya Islands, Sabah, Malaysia	197	516	647
Pulau Weh, Sumatra, Indonesia	196	533	644
Izu Islands, Japan	190	464	623
Christmas Island, Indian Ocean	185	560	606
Sipadan Island, Sabah, Malaysia	184	492	603
Rowley Shoals, Western Australia	176	505	576
Cocos-Keeling Atoll, Indian Ocean	167	528	545
North-West Cape, Western Australia	164	527	535
Tunku Abdul Rahman Is., Sabah, Malaysia	139	357	450
Lord Howe Island, Australia	139	395	450
Monte Bello Islands, W. Australia	119	447	382
Bintan Island, Indonesia	97	304	308
Kimberley Coast, Western Australia	89	367	281
Cassini Island, Western Australia	78	249	243
Johnston Island, Central Pacific	78	227	243
Midway Atoll	77	250	240
Rapa, Polynesia	77	209	240
Norfolk Island, Australia	72	220	223

lar short-term survey that is either currently in progress or already completed. For example, the CFDI for Milne Bay Province now stands at 337, and the appropriate regression formula (4.234 x 337 – 114.446) predicts an approximate total of 1,312 species, indicating that approximately 200 more species can be expected to occur in the province. On a much larger scale the CFDI can be used to estimate the reef fish fauna of the entire Indo-West Pacific region, a frequent subject of conjecture. This method estimates a faunal total of 3,764 species, a figure that compares favorably with the approximately 3,950 total proposed by Springer (1982). Moreover, Springer's figure covers shore fishes rather than reef fishes and thus include species not always associated with reefs (e.g., estuarine fishes).

The total CFDI for Milne Bay Province is 337 with the following components: Labridae (108), Pomacentridae (100), Chaetodontidae (42), Acanthuridae (34), Scaridae (28), and Pomacanthidae (25). This is the highest total for a restricted location thus far recorded in the Indo-Pacific, surpassing the previous figure of 333 for the Maumere Bay region of Flores, Indonesia. Table 4.6 presents a ranking of Indo-Pacific areas that have been surveyed to date based on

CFDI values. It also includes the number of reef fishes thus far recorded for each area, as well as the total fauna predicted by the CFDI regression formula.

Using CFDI values to compare more extensive regions, it can be seen from Table 4.7 that Papua New Guinea possesses the world's fourth richest reef fish fauna after Indonesia, Australia, and the Philippines. The number of actual species present in the Philippines remains problematical, but will no doubt exceed the total for Australia, which is the only country in the Coral Triangle region that has been adequately surveyed.

Zoogeographic affinities of Milne Bay Province fish fauna
Papua New Guinea belongs to the overall Indo-West Pacific faunal community. Its reef fishes are very similar to those inhabiting other areas within this vast region, stretching eastward from East Africa and the Red Sea to the islands of Micronesia and Polynesia. Although most families and many genera and species are consistently present across the region, the species composition varies greatly according to locality.

Milne Bay Province is an integral part of the Indo-Australian region, the richest faunal province on the globe in terms

Table 4.7. Coral fish diversity index (CFDI) for regions or countries with figures for total reef and shore fish fauna (if known), and estimated fauna from CFDI regression formula.

Locality	CFDI	No. reef fishes	Estimated
Indonesia	507	2060	2032
Australia (tropical)	401	1714	1584
Philippines	387	?	1525
Papua New Guinea	362	1494	1419
South Japanese Archipelago	348	1315	1359
Great Barrier Reef, Australia	343	1325	1338
Taiwan	319	1172	1237
Micronesia	315	1170	1220
New Caledonia	300	1097	1156
Sabah, Malaysia	274	840	1046
Northwest Shelf, Western Australia	273	932	1042
Mariana Islands	222	848	826
Marshall Islands	221	795	822
Ogasawara Islands, Japan	212	745	784
French Polynesia	205	730	754
Maldive Islands	219	894	813
Seychelles	188	765	682
Society Islands	160	560	563
Tuamotu Islands	144	389	496
Hawaiian Islands	121	435	398
Marquesas Islands	90	331	267

of biodiversity. The nucleus of this region, or Coral Triangle, is composed of Indonesia, Philippines, and Papua New Guinea. Species richness generally declines with increased distance from the coral Triangle. The damselfish family Pomacentridae provides an excellent, typical, example of the attenuation phenomenon. Indonesia has the world's highest total with 138 species, with the following totals recorded for other areas: Papua New Guinea/Papua Province (109), Northern Australia (95), Western Thailand (60), Fiji Islands (60), Maldives (43), Red Sea (34), Society Islands (30), and Hawaiian Islands (15).

Most of the fishes occurring in Milne Bay Province have relatively widespread distributions within the Indo-Pacific region. Nearly all coral reef fishes have a pelagic larval stage of variable duration, depending on the species. Therefore, the dispersal capabilities and length of larval life of a given species are usually reflected in the geographic distribution. Allen (1998) indicated that the largest zoogeographic category of Milne Bay fishes, containing about 30 percent of all species, includes widespread Indo-west and central Pacific forms, typically ranging from East Africa to Fiji or Samoa. Approximately 24 percent of the fauna ranges widely in the Indo-far western Pacific, and 18 percent are mainly western Pacific forms. The remaining species have more restricted distributions within the Indo-Pacific region. In addition, five species have worldwide distributions, occurring in all tropical seas.

Only five species are presently known only from Milne Bay Province: *Chrysiptera cymatilis* (Pomacentridae), *Cirrhilabrus pylei* (Labridae), *Novaculichthys* n. sp. (Labridae), *Ecsenius taeniatus* (Blenniidae), and *Trichonotus halstead* (Trichonotidae). However, future collecting will most likely expand their ranges to at least encompass neighboring provinces.

DISCUSSION AND RECOMMENDATIONS

Although a few segments of the marine biota, such as holothurians and giant clams, show signs of exploitation, most of the ichtyofauna is in excellent shape. Many parts of this vast province appear to be in pristine or near-pristine condition. It is therefore worthwhile to make comparison with more heavily exploited areas to the west where CI has conducted marine RAP surveys, prior to this one.

Table 4.8 presents the average number of species per site, the number of sites where more than 200 species were observed, and the greatest number seen at a single site for all Marine RAP surveys to date. The pristine nature of Milne Bay reef environments compared to other parts of the Coral Triangle is clearly evident from these statistics. A total of 200 or more species is generally considered the benchmark for an excellent fish count at a given site. This figure was obtained at 42 percent of Milne Bay sites, over twice as many times as its nearest rival, the Togean-Banggai Islands.

Milne Bay's dominance is further demonstrated in Tables 4.9 and 4.10. Table 4.9 ranks the top sites for fishes for the

Table 4.8. Comparison of site data for Marine RAP surveys 1997–2000.

Location	No. sites	Average spp./site	No. 200+ sites	Most spp. one site
Milne Bay, Papua New Guinea	110	192	46 (42%)	270
Togean/Banggai Islands, Sulawesi, Indonesia	47	173	9 (19%)	266
Calamianes Islands, Philippines	21	158	4 (10.5%)	208
Weh Island, Sumatra, Indonesia	38	138	0	186

five Indo-Pacific Marine RAPs between 1997–2000. Of the top 21 sites, 19 were located in Milne Bay Province.

Intra-provincial versus habitat variation

It was shown in the results section (see Table 4.5) that the Conflict Group and Nuakata region were the richest areas for reef fishes. Their dominance probably reflects a high percentage of outer-reef dropoffs. Exposed outer slopes consistently harbor a high number of species, although fringing reefs, which also feature a seaward slope component, are also comparatively rich. Ironically, both the highest and lowest species totals for a single site during the 2000 survey were from fringing reefs. This discrepancy underlines the importance of relative exposure. Highly sheltered fringing reefs with poor circulation and consequent siltation generally harbour fewer fishes than reefs exposed to oceanic surf and periodic strong currents.

Table 4.10 shows the correlation between species diversity and major habitats sampled during the 2000 survey. As mentioned in the previous paragraph, exposed outer reef slopes are the richest areas for fishes. This habitat was typical on the seaward edge of the Conflict Group and Louisiade Archipelago. Fringing reefs abound in Milne Bay Province, being the dominant reef type along the coastal mainland and around larger islands such as those found at the D'Entrecasteaux Group. Isolated platform reefs are also plentiful and typically rise steeply from deep water, either breaking the surface (at least at low tide) or form subtidal platforms. Keast Reef (site 19) was the best example of this type of structure encountered during the 2000 survey. Lagoon reefs supported the lowest fish diversity. This habitat was prevalent in the Louisiades Archipelago. It includes reefs surrounding individual islands of the Calvados Chain, which lie inside the huge lagoon formed by the outer barrier of the Louisiades. However, it could be argued they are actually fringing reefs.

Table 4.9. The 21 most speciose sites for Marine RAP surveys 1997–2000. (Total of 216 sites).

Rank	Location	Total fish spp.
1	Boirama Island, Milne Bay Province	270
2	East Irai Island, Milne Bay Province	268
3	Dondola Island, Togean Islands	266
4	Scub Islet, Sanaroa Island, Milne Bay Province.	260
5	Wahoo Reef, East Cape, Milne Bay Province	245
5	Mukawa, Cape Vogel, Milne Bay Province	245
7	Boia-boia Waga Island, East Cape, Milne Bay Province	243
8	West Irai Island, Conflict Group, Milne Bay Province	241
9	West barrier reef, Rossel Island, Milne Bay Province	238
10	Kathy's Corner, East Cape, Milne Bay Province	236
11	Gabugabutau, Conflicts, Milne Bay Province	235
12	Hardman Island, Laseinie Islands, Milne Bay Province	234
13	Tawal Reef, Louisiade Archipelago, Milne Bay Province	232
14	North tip of Unauna Island, Togean Islands	230
15	Swinger Opening, Rossel Island, Milne Bay Province	225
15	Punawan Island, Bramble Haven, Milne Bay Province	225
17	Southeast Butchart Island, Engineer Islands, Milne Bay Province	227
18	Catsarse Reef, Amphlett Islands, Milne Bay Province	224
18	Sabari Island, Louisiade Archipelago, Milne Bay Province	224
18	Rossel Passage, Milne Bay Province	224
20	Muniara Island, Conflict Group, Milne Bay Province	222

Table 4.10. Correlation between habitat type and average number of species per site.

Major habitat type	No. sites	Avg. no. spp./site
Outer reefs and passages	16	213
Fringing reefs	24	192
Isolated platform reefs	6	191
Lagoon reefs and patches	10	178

Rare or unusual species

Milne Bay Province is rapidly gaining the distinction of being the best-documented area of Papua New Guinea, and consequently it is extremely difficult to find new reef fishes. When I first dived in the Madang area in northern Papua New Guinea during 1972 there was a bonanza of undescribed taxa, even in relatively shallow depths. I have subsequently made numerous collecting trips to Papua New Guinea, invariably finding additional new species. Many of these were sent to specialist colleagues for further study, but

approximately 50 species were described by either Dr. Jack Randall (formerly of the Bishop Museum, Hawaii) or myself.

Only one new species was collected during the survey. It belongs to the damselfish (Pomacentridae), and was previously identified as *Pomacentrus smithi*, a species believed to be distributed in Papua New Guinea, Indonesia, and the Philippines (Allen, 1991). However, striking color differences are apparent between the Papua New Guinea fish and those seen on recent RAP surveys at Sulawesi and the Philippines. Photographs and a series of specimens were obtained during the Milne Bay RAP. Preliminary investigations reveal that two morphologically distinct species are represented. Compared to the Indonesian fish, the Milne Bay fish is distinguished by a yellow nape region, which has a pale gray nape. In addition, the Milne Bay fish has longer dorsal-fin spines and a deeper caudal-fin base. *Pomacentrus smithi* was originally described from the Philippines; therefore this name is applicable to the Indonesian and Philippines populations. The Milne Bay fish represents a new species. Specimens collected on past expeditions reveal that it also occurs at Madang, Manus Island, New Britain, Solomon Islands, and Vanuatu.

An unusual color variant of the pygmy angelfish *Centro-pyge bispinosa* was collected on outer reefs at the northwestern corner of the West Calvados Chain of the Louisiade Archipelago. Unlike the normal *C. bispinosa*, which have numerous, narrow blue bars on a reddish orange background, the Louisiade variants are characterized by a uniform whitish body or have faint brown spotting. This variation appears to be correlated with deeper water and was mainly confined to depths below about 15 m, down to at least 45 m. Normal *C. bispinosa* were found in less than 15 m, and a few "hybrid" specimens with an intermediate pattern were seen in the transition zone. Both normal fish and variants were very common, with hundreds of the deep-water fish being observed over mainly rubble bottoms. Normal-colored fish were observed in deep water at other outer reef sites in the Louisiade Archipelago, as well as other parts of Milne Bay Province.

Remarkable color variation was also noted in the damselfish *Acanthochromis polyacantha*, which until recently was believed to be the only member of this large (about 345 species) family to have non-pelagic larvae. Like most damselfishes, the parents guard the eggs, but instead of undergoing a pelagic stage at hatching, the young remain under their protection. Allen (1999) described two additional species in a new genus *Altrichthys*, which exhibits an identical life cycle. *Acanthochromis polyacantha* is monotypic, and over much of its range (especially the Philippines-Indonesian portion) exhibits a basic dark-brown color pattern. However, in the Australia-New Guinea region it shows a wide range of variation from almost entirely white to uniform dark brown, with other populations exhibiting variable amounts of these colors.

Although many subtle variations occur in *Acanthochromis* of Milne Bay Province, the two basic patterns are uniform dark brown and dark brown except for a white tail. A basic dark-brown pattern (often with narrow pale margins on the rear dorsal, anal, and caudal fins) is characteristic over the remainder of its range, which encompasses much of Indonesia and the Philippines. The dark fish was most common in Milne Bay Province, appearing at 43 sites, whereas the white-tailed variety was seen at only nine sites. In addition a uniform light gray variety was seen at sites 56 and 57. The detailed distribution of the two main types (dark tails and white tails) was fascinating. In some cases, for example between the Amphlett Islands (dark tail) and Sunday Island (white tail), the two varieties are separated by only 20 km. The white tails of Marx Reef (site 37) were separated from the dark tails of Rossel Passage (site 39) by a similar distance. Fish at the eastern end of Rossel lagoon had white tails, whereas those at the western end possessed dark tails. Detailed genetic studies of the New Guinean and Australian populations will probably reveal that numerous species are involved because the lack of a pelagic dispersal stage precludes the possibility of cross breeding between adjacent populations separated by deep oceanic water.

Conservation recommendations based on fish diversity

If reef fish diversity were the sole criteria for selecting priority areas for conservation purposes the areas listed below would certainly qualify. These recommendations are based on both the 1997 and 2000 surveys.

Nuakata area—This area centered on Nuakata Island contains numerous shoals or platform reefs. Most of these rise from deep water and come to within a few meters of the surface. They have exposed, steep seaward slopes and support a rich diversity of corals and fishes. The main island of Nuakata also supports a diverse fish assemblage, reflecting its habitat diversity, which includes mangroves, sheltered bays with fringing reefs, and more exposed seaward slopes. The small adjacent island of Boirama was the site of the highest fish count for all Marine RAPs to date (Table 4.8) and the second highest for all sites recorded by G. Allen over the past 30 years.

Conflict Group—This classic atoll is located midway between the Engineer Group and the Louisiade Archipelago. It supports the highest number of fishes per site of all areas surveyed to date in Milne Bay Province. The richest areas are on the outer, seaward slope of the small islets that encircle the central lagoon. The lagoon habitat, although not as diverse, contains an interesting fish community with a number of elements not seen on outer reefs.

Cape Vogel area—This region, marking the boundary between Collingwood and Goodenough bays, contains remarkable fish diversity in spite of occupying a relatively small area (roughly 100 km²). There are rich fringing reefs, mangrove shores, an extensive deep-water lagoon (between Ragrave and Sibiribiri points), and a small but very rich platform reef (Keast Reef). In addition, Kumbio Bay at the southern edge of Cape Vogel offers a highly sheltered fringing reef, which drops steeply from shore.

West Calvados Chain and Bramble Haven—Although we did not adequately survey this area due to adverse weather conditions, it holds promise as a very diverse location for fishes due to a good mix of habitats. It features small barrier reef islands, superb outer reef dropoffs, and abundant lagoon patch reefs. The lagoon reefs around Panasia (site 51) were very good for fishes. The single site (54) at Bramble Haven was one of the best for corals and fishes.

REFERENCES

Allen, G. R. 1991. Damselfishes of the World. Mentor, Ohio: Aquarium Systems.

Allen, G. R. 1993. Reef Fishes of New Guinea. Madang, Papua New Guinea: Christensen Research Institute.

Allen, G. R. 1998. Reef and shore fishes of Milne Bay Province, Papua New Guinea. *In*: Werner, T. B. and G. R. Allen (eds.). A rapid biodiversity assessment of the coral reefs of Milne Bay Province, Papua New Guinea. RAP Working Papers 11, Washington, DC: Conservation International. Pp. 39–49, 67–107.

Eschmeyer, R. N. (ed.). 1998. Catalog of Fishes, Vol. I–III. San Francisco: California Academy of Sciences.

Kuiter, R. H. 1992. Tropical Reef Fishes of the Western Pacific - Indonesia and Adjacent Waters. Jakarta: Percetakan PT Gramedia Pustaka Utama.

Myers, R. F. 1989. Micronesian reef fishes. Guam: Coral Graphics.

Randall, J. E., G. R. Allen, and R. C. Steene. 1990. Fishes of the Great Barrier Reef and Coral Sea. Bathurst, Australia: Crawford House Press.

Randall, J. E. and H. A. Randall. 1987. Annotated checklist of fishes of Enewetak Atoll and Other Marshall Islands. In: Devaney, D. M., E. S. Reese, B. L. Burch, and P. Helfrich (eds.). Vol 2. The natural history of Enewetak Atoll. Oak Ridge, Tennessee: Office of Scientific and Technological Information U.S. Dept. of Energy. Pp. 289–324.

Springer, V. G. 1982. Pacific plate biogeography with special reference to shorefishes. Smithsonian Contributions to Zoology 367: 1–182.

Winterbottom, R., A. R. Emery, and E. Holm. 1989. An annotated checklist of the fishes of the Chagos Archipelago, Central Indian Ocean. Royal Ontario Museum Life Sciences Contributions 145: 1–226.

Chapter 5

Living Coral Reef Resources of Milne Bay Province, Papua New Guinea

Mark Allen, Jeff Kinch, and Tim Werner

SUMMARY

- Preliminary stock abundance estimates for coral reef fishes, clams (Tridacnidae), and beche-de-mer (Holothuridae) were undertaken in Milne Bay Province, southeastern Papua New Guinea.

- A total of 209 species of fish, representing 69 genera and 27 families, were classified as target edible fishes. Fusiliers of the family Caesionidae were particularly abundant.

- Counts of target fish species for individual sites ranged between 27 and 56 (mean = 44.7 ± 1.21). Population counts of target fishes at each site ranged between 142 and 5,875 (mean = 801.4 ± 141.91).

- The vast majority of target fishes had an average size of less than 30 cm, with progressively greater numbers observed in decreasing size classes.

- The mean "site total" biomass estimate of target fishes is considerably higher in Milne Bay than at other surveyed areas within the "coral triangle," such as the Togean and Banggai Islands (Indonesia) and the Calamianes Islands (Philippines).

- The coral reefs of Milne Bay Province are fished relatively lightly by both commercial and artisanal sectors. Destructive fishing methods (e.g., dynamite and cyanide) have not been used in the area for 5 years, and reefs were in good to excellent condition at most sites.

- A total of 15 species of Holothuridae (sea cucumbers), representing four genera, were recorded from 53 sites. The most commonly observed and most abundant species were *Bohadschia argus*, *Pearsonothuria graeffei*, and *Thelenota anax*.

- The number of holothurian species recorded at each site ranged between 0 and 7 (mean = 2.6 ± 0.27). The number of individual holothurians recorded at each site ranged between 0 and 28 (mean = 6.0 ± 0.90).

- A total of six species of Tridacnidae (giant clams), representing two genera, were recorded from 39 sites. The most commonly observed and most abundant species were *Tridacna maxima*, *T. squamosa*, and *T. crocea*.

- The number of tridacnid species recorded at each site ranged between 0 and 4 (mean = 2.0 ± 0.16). The number of individual tridacnids recorded at each site ranged between 0 and 49 (mean = 9.4 ± 1.79).

- Stocks of Holothuridae and Tridacnidae seemed lower than might be expected, suggesting extensive harvesting. From these results a more thorough stock assessment involving CSIRO/NFA/CI was carried out.

- A summary of the utilization of living marine resources on Brooker Island, a remote community in the West Calvados Chain of the Louisiade Archipelago, is presented. The most intensively utilized fishery resources are beche-de-mer (Holothuridae), trochus, crayfish, fishes, and clams (Tridacnidae). During the period of January-September 1999, a total of 56,649.81 kina was earned through the sale of marine resources.

- Current and past exploitation of living marine resources in Milne Bay Province as shown particularly on Brooker Island, threaten the maintenance of commercially and biologically viable populations of target species.

INTRODUCTION

Milne Bay Province is dominated geographically by its marine environment, encompassing an estimated maritime area of 110,000 km² (Werner and Allen, 1998), a shoreline of 2,120 km, and over 600 islands, atolls, and offshore reefs (Omeri, 1991). Approximately 13,000 km² of coral reefs or about 32 percent of Papua New Guinea's total reef area (Munro, 1989) lies within the provincial boundary.

The bulk of Milne Bay Province's human inhabitants are situated near the seashore, both on the islands and the mainland. The sea and its resources play a vital role in the economy, livelihood, and customs of these people. Local communities are dependent on marine resources as a major source of nutrition and income. The impact exerted on the marine environment, through the harvesting of these resources, is likely to increase in the future given the burgeoning population of the province and the increasing need and desire for cash.

To ensure the sustainability of marine resources in Milne Bay Province, effective management strategies must be implemented. However, scientific data assessing the stocks of animals that are currently being exploited must be collected first. Previous stock assessments are now dated and have been limited to specific reef systems (see Chesher, 1980; Lindholm, 1978; Ledua et al., 1996) and based on export figures (Lokani and Ada, 1998) or inadequate sample numbers.

While these previous surveys provided some indicative abundance levels, species composition, and limited distribution data, they were not able to produce population parameter estimates useful for designing robust management strategies (see Kinch, 2002a; Skewes et al., 2002). Marine RAP surveys are aimed to provide a snapshot of fishing pressures in selected areas of the Province, which can suggest priorities for more detailed assessments.

A more thorough stock assessment was carried out during October and November, 2001, as a collaborative effort between the Commonwealth Scientific and Industrial Research Organisation (CSIRO), the National Fisheries Authority (NFA), and Conservation International (CI). The results of this survey are discussed below for Holothuridae and Tridacnidae.

A summary of the results of a socio-cultural study undertaken by J. Kinch (a former anthropology PhD student from the University of Queensland who is now employed by CI) of a community on Brooker Island in the Calvados Chain of the Louisiade Archipelago (Kinch, 1999) is also included and discussed below. The results of his research into marine resource utilization at the community level (gathered during 15 months in the field) represent an invaluable supplement to the RAP data and illustrate the kinds of human drivers of reef species exploitation in Milne Bay.

METHODS

The RAP survey area was located off the extreme southeastern tip of Papua New Guinea in Milne Bay Province. Fifty-seven sites were surveyed including localities in the D'Entrecasteaux and Amphlett islands, coastal areas of Collingwood and Goodenough bays, and in the Louisiade Archipelago.

Coral reef fishes

Data were collected visually while scuba diving and recorded with pencil on waterproof plastic paper. The visual census methodology outlined by Dartnall and Jones (1986) was employed with some modifications. The method entailed the placement of a tape measure (100 m in length) along the reef substrate by an assistant to the diver recording the census data. The recorder then moved slowly along the transect making observations for a distance of 10 m on either side of the tape, forming a survey area of approximately 2,000 m² per transect. The time spent on each transect ranged from 20 to 35 minutes. Data were collected for two transects at most sites: a deep one between 14 and 20 m and a shallow one at less than eight meters depth. Only one transect was completed for six of the sites due to either a lack of water depth, adverse local conditions, equipment malfunction, or medical problems with the divers.

Numbers of individuals and average length were recorded for every target species (see definition below) observed. Data for numbers of individuals were obtained by actual count except when fish occurred in large schools, in which case rough estimates were made to the nearest 50–100 fish. Average lengths were estimated to the nearest five centimeters.

These data were used to calculate fish biomass (expressed in ton/km²) following the methods of Sparre and Venema (1992). Average length was converted into weight using the cubic law:

Weight = 0.05(Length)³ (weight in grams and length in centimeters)

Target species

Target species are defined as edible fishes that live on or near coral reefs (La Tanda, 1998). It should be noted that fishers in Milne Bay Province do not currently exploit some of the target species that were recorded during the RAP. Nonetheless, these species were included in order to facilitate a meaningful comparison between the results of the present study and those of previous RAP surveys in other areas. The use of various field guides including Allen (1997), Allen and Swainston (1992), Lieske and Myers (1994), and Randall, Allen, and Steene (1990) aided the identification of target species.

Limitations of the study

The RAP took place during the month of June, a time of year characterized by strong winds and rough seas in Papua New Guinea's coastal zone, the severity of which usually forces local fishers to suspend fishing activities until conditions improve. In addition, the bad weather reduced the number of available survey sites, as intervening traveling times were much slower than anticipated.

Underwater visibility was also affected by the poor weather. In many lagoon areas visibility was reduced to less than 10 m. This made the visual census difficult, and may have contributed to a reduction in the accuracy of the quantitative data gathered. The cryptic habits of some target species (e.g., holocentrids and some serranids) undoubtedly had the same effect.

Some small initial-phase (ip) scarids (parrotfishes) could not be identified to species level with any certainty and were recorded simply as "ip scarid spp." Also, it was difficult to distinguish between some of the species belonging to the genus *Ctenochaetus* (Acanthuridae), so they were recorded as "*Ctenochaetus* spp." The species total recorded for each site could therefore be reduced by 1–3 species as identification was to only genus level. Also, note that data were recorded for only 50 (out of a possible 57) sites due to medical problems suffered by the fisheries consultant (MA), who could not dive for the last four days of the RAP.

Holothurians and tridacnids

Species diversity and abundance of commercial-sized holothurians (as defined for each species by The National Management Plan for Beche-de-mer Fisheries) and all tridacnids were recorded for each site. Size estimates of tridacnids were also recorded. Densities were reported when greater than 1 individual/m².

At each site, one or two divers descended to the base of the reef, usually in about 30 m or less, and then swam gradually up the reef slope to the surface, making observations mostly in the 0–15 meter depth range during an average dive time of 50 minutes. Most time was spent in shallower water as this is where these organisms are usually more abundant as well as being more accessible to local fishers.

Identifications were made using a combination of Gosliner *et al.* (1996), Colin and Arneson (1995), Allen and Steene (1994), and Conand (1998). All entirely black holothurians other than *Holothuria atra* were not identified to species, due to the limitation of available references.

RESULTS AND DISCUSSION

Coral reef fishes

A total of 209 target species, representing 69 genera and 27 families, were recorded (Appendix 6). Just over half of this total belonged to the families Serranidae, Acanthuridae, Scaridae, Lutjanidae, and Holocentridae.

The most commonly observed target species (percentage of occurrence in parentheses) were *Ctenochaetus* spp. (100%), *Scarus niger* (98%), *Monotaxis grandoculus* (96%), *Chlorurus bleekeri* (90%), *Cheilinus fasciatus* (88%), *Scarus flavipectoralis* (86%), *Naso lituratus* (80%), *Acanthurus pyroferus* (78%), and *Parupeneus barberinus* (78%). Appendix 6 lists the percentage of occurrences for all target species.

Caesionids, or fusiliers, were the most abundant fishes at most sites. Indeed, they comprised nearly 65 percent of the total "target" fish count (Fig. 5.1). Other prominent families contributing to the total count were Acanthuridae, Scaridae, Lutjanidae, and Carangidae. The most abundant species included *Pterocaesio pisang, Caesio cuning,* and *P. digramma* (Table 5.1).

Families that had less than five species observed over the course of the survey (i.e., those labeled as "others*" in Figs. 5.1 and 5.2) contributed the largest proportion of "target" fish biomass (Fig. 5.2), but only 1.8 percent of the total fish count (Fig. 5.1). This is simply a reflection of the larger average size (and thus biomass) of many species included in this group (e.g., sharks, rays, mackerels, barracudas).

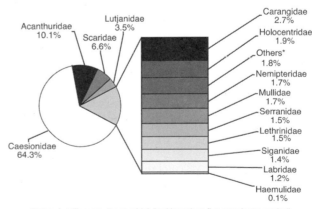

(*other families are those which had less than five species recorded)

Figure 5.1. Composition of total target fish count partitioned by family, Milne Bay RAP 2000.

Table 5.1. Summary table of coral reef fish stock assessment.

Site	Transect 1 (14–20m)		Transect 2 (< 8m)		Site total		Most abundant species (% of total fish count)
	No. target species	Approx. fish count	No. target species	Approx. fish count	No. target species	Approx. fish count	
1	36	204	38	490	55	694	*Pterocaesio pisang* (28.82%)
2	40	807	-----	-----	40	807	*Caesio cuning* (37.17%)
3	-----	-----	30	208	30	208	*Pterocaesio pisang* (48.08%)
4	33	449	33	655	48	1,104	*Pterocaesio pisang* (54.35%)
5	33	1,392	-----	-----	33	1,392	*Pterocaesio pisang* (43.10%)
6	42	1,733	32	1,714	55	3,447	*Gymnocaesio gymnopterus* (58.02%)
7	31	270	27	135	42	405	*Caesio caerulaurea* (22.22%)
8	26	162	36	116	48	278	*Pterocaesio digramma* (17.99%)
9	42	247	37	1,301	55	1,548	*Gymnocaesio gymnopterus* (25.84%)
10	30	233	33	440	48	673	*Pterocaesio pisang* (52.01%)
11	20	171	35	267	40	438	*Pterocaesio pisang* (45.66%)
12	33	462	38	225	48	687	*Pterocaesio pisang* (46.58%)
13	32	184	34	283	46	467	*Caesio cuning* (51.39%)
14	-----	-----	25	150	25	150	*Ctenochaetus* spp. (28.00%)
15	28	208	25	151	43	359	*Caesio cuning* (33.43%)
16	32	219	42	237	52	456	*Caesio cuning* (18.42%)
17	34	288	32	269	47	557	*Caesio cuning* (39.50%)
18	16	342	36	392	42	734	*Pterocaesio pisang* (40.87%)
19	13	253	26	214	32	467	*Pterocaesio pisang* (51.39%)
20	29	226	26	390	36	616	*Pterocaesio pisang* (55.19%)
21	-----	-----	23	80	23	80	*Ctenochaetus* spp. (23.75%)
22	30	243	38	153	49	396	*Caesio cuning* (24.75%)
23	18	150	30	162	35	312	*Pterocaesio pisang* (35.26%)
24	21	145	17	181	29	326	*Pterocaesio pisang* (39.88%)
25	29	243	35	246	42	489	*Caesio cuning* (21.68%)
26	18	62	22	80	28	142	*Ctenochaetus* spp. (21.13%)
27	20	86	35	179	40	265	*Caesio lunaris* (22.64%)
28	37	605	35	510	50	1,115	*Caesio cuning* (31.39%)
29	31	486	37	131	49	617	*Pterocaesio trilineata* (40.52%)
30	28	263	40	328	48	591	*Pterocaesio digramma* (27.07%)
31	21	118	33	218	39	336	*Ctenochaetus* spp. (21.49%)
32	36	460	31	245	45	705	*Pterocaesio digramma* (26.24%)
33	39	512	32	193	50	705	*Pterocaesio pisang* (42.55%)
34	34	249	37	389	53	638	*Pterocaesio digramma* (18.81%)
35	33	175	30	133	48	308	*Pterocaesio digramma* (19.48%)
36	34	399	39	308	53	707	*Caesio cuning* (38.19%)
37	39	582	42	554	54	1,136	*Pterocaesio pisang* (33.45%)
38	21	186	32	160	39	346	*Plotosus lineatus* (23.12%)

continued

Table 5.1. Summary table of coral reef fish stock assessment *(continued).*

Site	Transect 1 (14–20m) No. target species	Transect 1 (14–20m) Approx. fish count	Transect 2 (< 8m) No. target species	Transect 2 (< 8m) Approx. fish count	Site total No. target species	Site total Approx. fish count	Most abundant species (% of total fish count)
39	26	141	44	333	52	474	*Pterocaesio pisang* (25.32%)
40	-----	-----	27	138	27	138	*Plotosus lineatus* (36.23%)
41	22	125	20	349	30	474	*Amblygaster sirm* (31.65%)
42	21	514	33	558	41	1,072	*Pterocaesio pisang* (46.64%)
43	32	280	38	198	46	478	*Pterocaesio digramma* (23.01%)
44	35	3,136	43	2,739	53	5,875	*Gymnocaesio gymnopterus* (59.57%)
45	40	174	43	214	56	388	*Ctenochaetus* spp. (15.21%)
46	30	721	32	613	47	1,334	*Pterocaesio pisang* (63.72%)
47	29	247	42	385	55	632	Unidentified IP Scarids (23.73%)
48	17	418	21	327	30	745	*Pterocaesio tile* (53.69%)
49	22	1,029	19	81	35	1,110	*Pterocaesio pisang* (45.05%)
50	25	424	14	190	33	614	*Lutjanus kasmira* (48.86%)
n	46	46	48	48	44*	44*	
Minimum	13	62	14	80	27	142	**Ranking of most abundant species**
Maximum	42	3,136	44	2,739	56	5,875	1. *Pterocaesio pisang* (19 sites)
Average	29.09	435.28	32.27	375.25	44.68	801.36	2. *Caesio cuning* (9 sites)
Stand. Err.	1.10	76.81	1.05	65.48	1.21	141.91	3. *Pterocaesio digramma* (5 sites)

* n for the site totals is 44 as sites where only one transect was surveyed have been ignored.

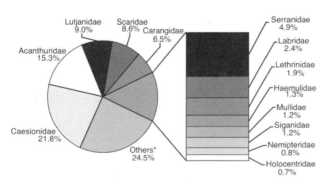

(*other families are those that had less than five species recorded)

Figure 5.2. Composition of total target fish biomass partitioned by family, Milne Bay RAP 2000.

Summary of data for sites (Table 5.1)
Counts of target species ranged from 13 to 42 (mean = 29.1 ± 1.10) on deep transects and from 14 to 44 (mean = 32.3 ± 1.05) on shallow transects. An unpaired student's t-test revealed a significant difference between the mean values at a 5 percent level of significance (Prob. (2-tail): 0.039, df: 92). The target species count for both transects combined (i.e., the site total) ranged from 27 to 56 (mean = 44.7 ± 1.21).

Numbers of individual target fish counted ranged from 62 to 3,136 (mean = 435.3 ± 76.81) on deep transects and from 80 to 2,739 (mean = 375.2 ± 65.48) on shallow transects with no significant difference between depths (unpaired t-test: t = 0.552, p > 0.05, df: 92). Fish counts for both transects combined ranged from 142 to 5,875 (mean = 801.4 ± 141.91). On average the majority of observed target fishes were less than 30 cm in length, with progressively greater numbers recorded in decreasing size classes (Fig. 5.3). Fish belonging to the larger size classes were typically not abundant.

Intra-provincial comparison of coral reef fishes data
Sites were grouped by geographic area and habitat type (Table 5.2), to elucidate any trends in the data (Table 5.3). A few trends may be evident, however a lack of repetition

within some of the geographic and habitat groups precluded the use of statistics to test for differences in their mean values. Nonetheless, the platform/patch reef and outer reef/passage sites appear to have a larger target fish count compared to the fringing and lagoonal reef sites (Table 5.4). This apparent trend may be at least partly attributable to the fact that artisanal fishers have limited access to platform and outer reefs compared to fringing and lagoonal reefs that are more sheltered and closer to villages.

Comparison of Milne Bay Province with other areas
The mean "site total" estimate of target fish biomass is considerably higher for Milne Bay Province than for both the Togean-Banggai Islands of Indonesia, and the Calamianes Islands of the Philippines (Fig. 5.4). This is a strong

reflection of the relatively pristine condition of coral reefs throughout Milne Bay Province, and also highlights the disparity that exists in levels of exploitation of reef fish stocks in different regions throughout the "coral triangle."

Coral reefs of Milne Bay Province are fished relatively lightly by both the commercial and artisanal sectors, as compared with the generally more exploited reefs of Indonesia and the Philippines. Human population density, and consequent artisanal fishing pressure, in Milne Bay Province is substantially lower than other areas where RAP surveys have taken place (G. Allen, pers. comm.), which helps to explain the higher reef fish biomass encountered here. Additionally, destructive fishing methods (e.g., dynamite, cyanide) have not been used in Milne Bay Province for five years. In the other places where RAP surveys have taken place, such practices are considerably more common.

Conclusions from coral reef fishes data
Coral reefs of Milne Bay Province support a rich diversity and abundance of fishes. Reefs were in good to excellent condition at most sites (chapter 2, this report), with little evidence of negative environmental impact by humans.

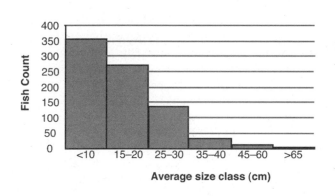

Figure 5.3. Number of target fish belonging to different size classes at an average site*, Milne Bay RAP 2000 (*excluding sites where only one transect was surveyed).

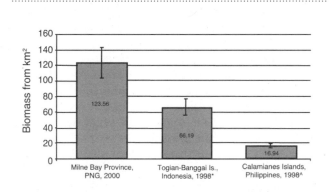

Figure 5.4. Comparison of mean "site total" biomass of target fishes for past and present CI RAP surveys (sources: * La Tanda, 1998; ^ Ingles, 1998).

Table 5.2. Geographic and habitat groupings used for data (fisheries data were not available for all sites).

Geographic area groups	Sites
1. North Coast	1, 14–28
2. D'Entrecasteaux Islands	2–5, 13
3. Amphlett Islands	6–12
4. Louisiade Arch./Conflict Group	29–54
5. South-east Islands	55–57
Habitat type groups	**Sites**
1. Fringing reef	1, 3, 4, 7–8, 10–15, 17–18, 20, 22–28, 55–57
2. Platform/patch reef	2, 5, 6, 9, 16, 19
3. Outer (barrier) reef or passage	29–30, 35, 37–39, 43–45, 47–50, 52–54
4. Lagoonal reef	21, 31–34, 36, 40–42, 46, 51

Table 5.3. Summary of target fish data grouped by geographic area.

Geographic group	n	Mean values for "site total" (± standard error)	
		No. target species	Approx. fish count
North Coast	14	41.4 (± 2.3)	494.9 (± 65.0)
D'Entrecasteaux Is.	2	47.0 (± 1.0)	785.5 (± 318.5)
Amphlett Is.	7	48.0 (± 2.2)	1068.0 (± 426.8)
Louisiade Arch./ Conflict Group	21	45.5 (± 1.8)	918.3 (± 255.3)

Table 5.4. Summary of target fish data grouped by habitat type.

Habitat type group	n	Mean values for "site total" (± standard error)	
		No. target species	Approx. fish count
Fringing reef	19	42.9 (± 1.6)	529.3 (± 60.3)
Platform/patch reef	4	48.5 (± 5.5)	1479.5 (± 704.1)
Outer (barrier) reef or passage	13	46.0 (± 2.5)	1024.2 (± 410.5)
Lagoonal reef	8	44.8 (± 2.8)	746.4 (± 112.7)

Moreover, Milne Bay Province compares very favorably with other areas that Conservation International has assessed in recent years.

Milne Bay Province has excellent potential for sustainable marine based eco-tourism, particularly diving. This industry is definitely compatible with artisanal and commercial fisheries at current operating levels. Certainly one of the more appealing aspects of the marine environment in Milne Bay Province is the general absence of destructive fishing methods that plague the more populated areas of Southeast Asia such as Indonesia and the Philippines. Efforts should be made to reinforce the value of sustainable fisheries management among the artisanal fishing sector of Milne Bay Province, and to ensure that destructive fishing methods do not become a temptation in the future.

Holothurians and Tridacnids
Holothuridae (beche-de-mer)
During 2002, 19 species of holothurians, most of which were of low economic value, were fished commercially from Milne Bay waters (Kinch, 2002a). The beche-de-mer fishery in Milne Bay Province is changing from a low-volume, high-value trade to a high-volume, low-value one. During the 2000 RAP survey, a total of 15 species representing five genera were recorded from 53 sites (Table 5.5). The most commonly observed species (percentage of occurrence in parentheses) were *Bohadschia argus* (43.40%), *Pearsonothuria graeffei* (39.62%), *Thelenota anax* (30.19%), *Holothuria atra* (24.53%), and Stichopus variegatus (24.53%). The most abundant species were *Thelenota anax, Bohadschia argus,* and *Pearsonothuria graeffei,* whose combined numbers comprised roughly half of the total count (Table 5.5).

The number of holothurian species recorded at a site ranged between 0 and 7 (mean = 2.6 ± 0.27). The number of individual holothurians recorded at a site ranged between 0 and 28 (mean = 6.0 ± 0.90). These animals were most frequently observed on sandy or rubble-covered slopes, and very occasionally among live coral. At most sites the density of holothurians was sparse (< 1 individual/m²), but occa-

sional isolated patches with higher population densities (up to 6 individuals/m²) were encountered.

The depth distribution of holothurians was fairly even, however, a slightly higher number was recorded from depths between 5 and 9 m (Fig. 5.5). Holothurians are typically most abundant in shallower water (0-10 m depth), but our data do not entirely support this trend. These results possibly point to the impact that local beche-de-mer fishers are having on stocks in shallow water.

Fishers generally collect these animals from depths of less than 15–25 m, as diving methods are limited to breath holding, although some illegal use of scuba and hookah equipment has been reported (see Kinch, 2002a). Another fishing technique involving the use of hooks attached to lead weights is sometimes employed to harvest beche-de-mer from deeper water.

Intra-provincial comparison of holothurian data
Site surveys for holothurian data were not replicated sufficiently within geographical groupings to conform with standards of statistical validity. However, counts of individual holothurians appeared to be higher at sites located in the South-east Islands (Table 5.6). Also, species diversity of holothurians appeared to be greater on fringing and platform reefs than on outer and lagoonal reef types (Table 5.7).

The RAP survey stock estimates indicated some depletion of the commercial holothurians, especially the higher value species. These results provided an impetus for a more detailed assessment of the marine resources of Milne Bay Province that was undertaken during October and November, 2001, as a collaborative effort between CSIRO, NAF, and CI.

The specific objectives of the stock assessment were to determine habitat condition and the state of holothurian and other benthic resources (e.g., clams) by surveying 1,126 sites within Milne Bay Province. Abundance, distribution, size frequency, and biological data were gathered and used to calculate stock size estimates for the area with sufficient precision to be useful for formulating sustainable management strategies for each of the commercially important holothurian species. The ultimate objective was to recommend and implement these management strategies for the sustainable exploitation of beche-de-mer in Milne Bay Province (CSIRO, 2001; Kinch, 2002a; Kinch *et al.*, 2001a; Skewes *et al.*, 2002).

The mean density for all commercial species was 21.24/ha. Mean density for high-value species was 5.22/ha. Individual species had mean densities as follows: *Holothuria nobilis* at 0.18/ha, *H. scabra* at 0.00/ha, *H. fuscogilva* at 0.42/ha, *H. edulis* at 2.15/ha, *H. atra* at 9.81/ha, *H. fuscopunctata* at 0.04/ha, *B. argus* at 1.33/ha, *B. vitiensis* at 0.99/ha, *Pearsonothuria graeffei* at 0.37/ha, *Stichopus chloronotus* at 3.81/ha, *S. variegatus* at 0.09/ha, *Thelenota ananas* at 0.47/ha, *T. anax* at 0.63/ha, *Actinopyga miliaris* at 0.12/ha, *A. lecanora* at 0.02/ha, and *A. mauritiana* at 0.12/ha (Skewes *et al.*, 2002; Kinch, 2002a).

Table 5.5. List of holothurian species recorded from Milne Bay Province (2000).

Species	No. sites where present	Percent occurrence	Total number of individuals recorded	Percent of total count
Bohadschia argus	23	43.40%	49	15.31%
Bohadschia vitiensis	4	7.55%	5	1.56%
Holothuria atra	13	24.53%	31	9.69%
Holothuria edulis	11	20.75%	24	7.50%
Holothuria fuscogilva	9	16.98%	14	4.38%
Holothuria fuscopunctata	7	13.21%	17	5.31%
Holothuria nobilis	3	5.66%	4	1.25%
Pearsonothuria graeffei	21	39.62%	40	12.50%
Stichopus chloronotus	3	5.66%	4	1.25%
Stichopus horrens	1	1.89%	1	0.31%
Stichopus variegatus	13	24.53%	19	5.94%
Stichopus spp.	1	1.89%	1	0.31%
Thelenota ananas	10	18.87%	22	6.88%
Thelenota anax	16	30.19%	70	21.88%
Thelenota rubralineata	5	9.43%	19	5.94%
		Total	**320**	**100.00%**

Table 5.6. Summary of holothurian data for geographic area groups.

Geographic group	n	Mean values for "site total" (± standard error)	
		No. species	Holothurian count
North Coast	14	3.1 (± 0.63)	7.5 (± 2.08)
D'Entrecasteaux Is.	4	3.5 (± 1.50)	5.2 (± 2.17)
Amphlett Is.	6	3.2 (± 0.75)	7.5 (± 2.42)
Louisiade Arch./ Conflict Group	26	1.9 (± 0.29)	4.3 (± 1.14)
South-east Islands	3	4.3 (± 0.88)	12.3 (± 4.63)

Table 5.7. Summary of holothurian data for habitat type.

Habitat type group	n	Mean values for "site total" (± standard error)	
		No. species	Holothurian count
Fringing reef	21	3.5 (± 0.49)	8.3 (± 1.67)
Platform/patch reef	5	3.0 (± 1.05)	5.8 (± 1.39)
Outer (barrier) reef or passage	16	1.6 (± 0.36)	4.3 (± 1.73)
Lagoonal reef	11	2.5 (± 0.39)	4.5 (± 1.07)

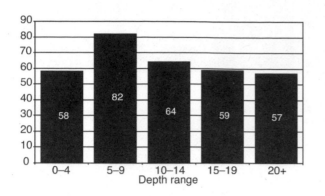

Figure 5.5. Counts of holothurians at different depth ranges, Milne Bay RAP 2000.

Results from the CSIRO/NFA/CI stock assessment show that holothurian stocks within Milne Bay Province are facing serious decline. The current Total Allowable Catch (TAC) of 140 metric tonnes is above the estimated maximum sustainable yield of 108 metric tones (Skewes *et al.*, 2002). *H. scabra* was not observed during the survey, *H. nobilis* was also found in very low numbers, and *H. fuscogilva* and *T. ananas* were showing signs of harvesting pressure. It has been recommended that the fisheries for *H. scabra* and *H. nobilis* be closed or have species specific TACs. Given the recent history of TAC levels being exceeded in Milne Bay Province, the catch should be monitored carefully and TAC strictly enforced. Individual local level government catch limits, the enforcement of current minimum size limits, and education of fishers to return undersize animals alive, are other mechanisms that need to be considered for the sustainable management of the Milne Bay fishery (Skewes *et al.*, 2002).

Further work is still required if Milne Bay Province holothurian fishery is to remain viable. There have been neither surveys during the closed fishing season nor prior to its opening as a means of assessing inter-annual changes in stock variability. Based on current information from the CSIRO/NFA/CI stock assessment, there is definite evidence of over-fishing in certain areas of Milne Bay Province, and the TAC now requires review. The CSIRO/NFA/CI stock assessment allows for repeated-measures sample strategies that could allow for a very efficient fishery independent monitoring program to modify the TAC as necessary and monitor population trends (see CSIRO, 2001). It is also important to establish sustainable management systems with supporting policy incentives to ensure that commercially valuable species do not become extinct. Steps now need to be undertaken to limit the effort exerted on the stocks as a loss of income and depletion of future stocks through the indiscriminate collection and subsequent rejection of undersized holothurians will cause dire social problems.

Management strategies that could be tested include having TACs set at the local level government (LLG), and for certain species with low abundances it maybe necessary to implement specific closures by setting species specific TACs. Resources need to be allocated for awareness and capacity building at the village level for management of these valuable resources. These would include extension and training materials on processing and appropriate harvesting methods; village awareness of over-fishing on resource sustainability; and the possible incorporation of traditional closed seasons or areas (the best means of policing closed areas may be through village involvement) and limited entry. Effective monitoring is necessary to prevent over-exploitation and depletion of holothurian resources. Further study is required on models of resource extraction. There is a need to continue stock assessments; monitor active fisheries and recovery rates; apply proper enforcement of data recording; and provide empowerment and support for fisheries inspectors and monitoring of overseas markets. Finally, the potential for hatchery and re-seeding programs should be investigated. Undoubtedly, there are major requirements for immediate reform in order to establish a sustainable fishery and improve upon this important industry.

Tridacnidae (clams)

A total of six species representing two genera were recorded from 39 sites (Table 5.8). The most commonly observed species (percentage of occurrence in all sites in parentheses) were *Tridacna maxima* (69.23%), *T. squamosa* (56.41%), and *T. crocea* (41.03%). The same three species were also the most numerous, combining to make up over 90 percent of the total count. The most abundant tridacnid species were also the smallest in size (Table 5.8). The fact that the medium-to-large species made up such a small percentage of the overall count suggests that local fishers are having an impact on their numbers. However, smaller clams are actually more highly prized by the local people as they are considered better eating. Moreover, the trend is obviously a feature of the ecology of these animals, with the smaller species being naturally more abundant.

The number of tridacnid species recorded at each site ranged between 0 and 4 (mean = 2.0 ± 0.16). The number of individuals recorded at each site ranged between 0 and 49 (mean = 9.4 ± 1.79). Densities were low at most sites (i.e., less than 1 individual/m²), but occasionally isolated patches with higher densities (up to 9 individuals/m²) were encountered. The vast majority of tridacnids were recorded at depths between 0 and 9 m (Fig. 5.6).

Intra-provincial comparison of tridacnid data

Despite the fact that low repetition rendered statistical analysis impractical, there appears to be a greater mean count of tridacnids at sites in the Louisiades/Conflict geographic group (Table 5.9).

Lack of replication also limited the statistical comparisons of tridacnid data between the various habitat type treatments. The platform/patch reef group, for example, was not represented (Table 5.10). Our data suggest that tridacnid

Table 5.8. List of tridacnid species recorded during the RAP survey.

Species	No. sites where present	Percent Occurrence	Total number of individuals recorded	Percent of total count	Average length (cm)
Tridacna crocea	16	41.03%	108	29.43%	6.6 (± 0.37)
Tridacna derasa	3	7.69%	8	2.18%	50.3 (± 5.01)
Tridacna gigas	7	17.95%	13	3.54%	81.8 (± 7.43)
Tridacna maxima	27	69.23%	182	49.59%	13.9 (± 0.81)
Tridacna squamosa	22	56.41%	52	14.17%	20.9 (± 2.11)
Hippopus hippopus	4	10.26%	4	1.09%	30.5 (± 8.63)
		Total	**367**	**100.00%**	

Table 5.9. Summary of tridacnid data for geographic area groups.

Geographic group	n	Mean values for "site total" (± standard error)	
		No. species	Tridacnid count
North Coast	5	2.6 (± 0.24)	7.6 (± 1.91)
D'Entrecasteaux Is.	2	2.5 (± 1.50)	4.0 (± 2.00)
Amphlett Is.	4	1.5 (± 0.29)	1.8 (± 0.48)
Louisiade Arch./ Conflict Group	25	1.8 (± 0.20)	11.6 (± 2.64)
South-east Islands	3	3.0 (± 0.58)	8.00 (± 3.21)

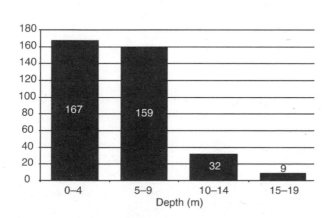

Figure 5.6. Counts of tridacnids at different depth ranges, Milne Bay RAP 2000.

counts may be higher than observed on lagoonal reefs (Table 5.10).

The first stock abundances estimated for tridacnid species were carried out in 1980 by Chesher in the southeast of the province. He stated that prior to commercial harvesting, unfished areas of southern Milne Bay Province contained an overall density of 39/ha for all species of giant clams (Chesher, 1980). In 1996, a stock assessment by the South Pacific Commission (SPC) and the NFA was carried out in the Engineer and Conflict Groups of islands. Throughout the survey area *Hippopus hippopus* was the most abundant at 20.1/ha, followed by *Tridacna maxima* at 17.9/ha, *T. crocea* at 11.9/ha, *T. squamosa* at 5.8/ha, *T. derasa* at 5.3/ha, *T. gigas* at 0.4/ha, and *Hippopus porcellanus* at 0.3/ha. From extrapolation of these data it was suggested by Ledua *et al.* (1996) that approximately 98 percent of the stock of *Tridacna gigas* throughout the province had been wiped out since the early 1980s, with overall stock density of all species estimated to be down by 82.35 percent of the original population. The overall density of all species was estimated to be only 0.5/ha (Ledua *et al.*, 1996).

Given the importance of giant clams to the people of Milne Bay Province, and following the results of the 2000 RAP, a more thorough and comprehensive assessment of tridacnid stocks was required. The CSIRO/NFA/CI stock assessment referred to previously in this chapter also included an appraisal of giant clam stocks. Abundances of *Tridacna gigas* ranged from 0 to 1.32/ha throughout the Province with a mean density of 0.82/ha. Mean densities for the other species were *Tridacna maxima* at 1.79/ha, *T. derasa* at 0.34/ha, *T. squamosa* at 1.37/ha, *T. crocea* at 14.85/ha, and *Hippopus hippopus* at 0.41/ha. The species harvested for commercial use had a mean density of 4.32/ha (Kinch, 2002c). The results from the CSIRO/NFA/CI stock assessment are consistent with other reports that indicate that stock levels are very low and have been heavily depleted across Milne Bay Province.

In order to ensure the sustainability of all marine resources in Milne Bay Province, effective management strategies must

Table 5.10. Summary of tridacnid data for habitat type.

Habitat type group	n	Mean values for "site total" (± standard error)	
		No. species	Tridacnid count
Fringing reef	14	2.4 (± 0.27)	5.5 (± 1.17)
Platform/patch reef	0	-----	-----
Outer (barrier) reef or passage	15	1.7 (± 0.27)	9.0 (± 2.40)
Lagoonal reef	8	2.0 (± 0.33)	17.1 (± 6.66)

be implemented. Several approaches to assist the recovery of overfished tridacnid populations have been proposed. These include establishing Locally Managed Marine Areas (LMMA), concentrating the remaining adult clams so that their reproduction can be facilitated by their closer proximity; and seeding cultured giant clams of sufficient size or in sufficient numbers and releasing these into the field to produce adult populations (Kinch, 2001b and 2002c). The imposition of either a ban on further commercial fishing or strict harvesting quotas, such as a single annual short harvesting season, could also offer prospects for management.

Conclusions from holothurians and tridacnids data

Milne Bay Province is the largest producer of holothurians in PNG, and previously the only province exporting tridacnid products. The income derived by coastal communities, particularly from the sale of holothurians in Milne Bay Province has increased dramatically throughout the 1990s and communities are currently landing large amounts. The increase in production of holothurians can be attributed to the decline in copra prices, the effects of drought in previous years, increased fishing for holothurians in remote locations, a decline in the value of other marine resources, and the establishment of new markets for previously low-value or non-commercial species. Subsequently, the diversity of holothurians is now being altered in some areas due to this intensified and extensive exploitation. This represents a threat to community livelihood strategies, the fishery itself, and the overall biodiversity of Milne Bay Province's reef ecosystems. Tridacnid exports and commercial exploitation have now been stopped under a ruling of the Department of Environment and Conservation (see Kinch, 2001b and 2002c).

As suggested by the RAP survey and substantiated by the CSIRO/NFA/CI stock assessment, stocks of holothurians and tridacnids are lower than normal. Consequently, fisheries need to adopt sustainable models of development if they are to provide maximum long-term yields. Besides contributing to the overall protection of Milne Bay Province's unique and valuable marine biodiversity, sustainable fisheries will allow villagers to maintain and even increase their incomes, thus making them less likely to migrate to urban areas in search of better employment. Sustainable management systems that are supported by strategic policy incentives need to be established soon to ensure that commercially valuable species do not become extinct.

Marine Resource Utilization on Brooker Island

Brooker Island is located in the West Calvados Chain of the Louisiade Archipelago. It is home to a small island community numbering just over 400 people, and the population is increasing at roughly 4.3 percent annually. The community at Brooker is largely dependent on the marine environment as they only produce approximately 50 percent of their own subsistence requirements from agriculture. They are thus avid sailors and major marine resource exploiters (Kinch, 1999; 2001a).

The ocean and its resources are critically important to these people (and indeed those throughout coastal PNG). Their livelihood and entire way of life is dependent on the exploitation of the marine environment. In order to maintain long-term, sustainable use of these resources, exploitation pressure must be regulated in such a way as to allow the environment to maintain its stability and regenerative capacity. This is becoming increasingly difficult for many communities throughout PNG, including Brooker.

The people of Brooker Island utilize approximately 5,000 km of sea territory (a proportion of this territory was under dispute with another island community) for the procurement of marine resources, and this area encompasses an extensive and diverse marine environment. From July 1998 to June 1999, the Brooker Community earned 67,000 kina from the sale of various marine resources, and this accounted for approximately 90 percent of all income. Of this 67,000 kina, beche-de-mer contributed 49.3 percent to the total; trochus, 19 percent; crayfish, 13.1 percent; fish, 10.8 percent; giant clams, 6.7 percent; and the remainder from shark fin and black lip pearl oyster (Kinch, 1999, 2001a). An account of the utilization of resources follows, separated into sections according to the type of animal exploited (turtles, sharks, beche-de-mer, giant clams and other shells, fishes and crayfishes).

Turtles

Marine turtles are heavily utilized as a food source, for trading, and to a limited extent for sale in local markets. Turtles found in Brooker waters include the hawksbill *(Eretmochelys imbricata)*, green *(Chelonia mydas)*, loggerhead *(Caretta caretta)*, and leatherback *(Dermochelys coriacea)*, of which the hawksbill and green are the most heavily utilized.

The turtle season begins approximately in October and ends in April. This corresponds with a time of year when very little food is available from local gardens. During the 1998–99 nesting season, a total of 149 green turtles and 50 hawksbill turtles were harvested. Turtle eggs are also highly prized as a food source, and eggs from a total of 604 nests belonging to these two species were gathered from 26 different nesting sites during the season.

Locals generally concur that turtle numbers have been fluctuating during the past few years. The *El Niño* climatic phenomenon has been implicated in this fluctuation, due to the regulating effect it has on seagrass nutrients and subsequent nutritional status and life history of turtles (Lanyon *et al.*, 1989). Turtle poaching is a problem in some areas, and there have been reported cases of the slaughter of turtles that are left to rot without the meat being consumed or taken to market (Kinch, 1999, 2001a).

The Milne Bay Provincial Government has, in the past, broadcasted a turtle awareness program on local radio stations. Education programs such as this are important if marine turtles are to be successfully conserved in the region. CI has developed materials to assist in education and awareness (Kinch, 2002b) and is currently developing a turtle watch and monitoring program with the Environment Protection Agency in Australia and the South Pacific Regional Environment Program (Kinch, 2002d). Local residents of Milne Bay Province must become more active in protecting and maintaining turtle habitats. A detailed study of marine turtle populations and seasonal abundance in Milne Bay Province is needed to accurately assess the status of these animals here

Sharks
During the period of September 1998 to October 1999, sharks comprised a minor part of the overall fish harvest on Brooker Island. Two specialist hunters were active at this time on the island. During a recorded two-month period (June–July 1999), one of these hunters caught a total of 39 sharks. This catch included black-tip reef sharks (*Carcharhinus melanopterus)*, lemon sharks *(Negaprion acutidens)*, white-tip reef sharks (*Triaenodon obesus*), gray reef sharks (*Carcharhinus amblyrhynchos*), tiger sharks *(Galeocerdo cuvier)*, hammerheads (*Sphyrna* spp.), and various other unidentified sharks.

Beche-de-mer (holothurians)
Beche-de-mer is a valuable coastal fishery in PNG and is exclusively export orientated. The beche-de-mer artisanal fishery involves coastal and island communities of fishermen, buyers who buy the processed holothurian products from the fishermen, and exporters (both licensed and illegal) who export the processed holothurians to the international market. Most exporters and buyers are based in Alotau and use diesel powered boats to visit the villages on purchasing trips. Milne Bay Province has seen a rise in the contribution it makes to the total PNG exports. In the early to mid-1990s this percentage fluctuated between 10 and 15 percent, but has seen an exponential growth, supplying just under half in 2001 (Kinch, 2002a). The large increase in beche-de-mer production resulted from circumstances common to all coastal areas in the province, including the consequences of drought, the decline in copra prices, and decreasing value of other marine species harvested by villagers. The fishery plays

an important role in maintaining rural social stability by providing income-earning opportunities in remote locations where opportunities are limited.

From 5 January to 1 May 1999, J. Kinch (in prep.) recorded 121 trips where fishers from Brooker Island were targeting giant clams, holothurians and crayfish in the Long/Kosmann Reef area surrounding Nagobi and Nabaina Islands, and the Bramble Haven Group. These trips were divided into three sub-types depending on use of vessels and main targeted species. These include:

- Trip type 1: Fishers harvesting holothurians as the main target species, with giant clams taken opportunistically. Fishers operating from sailing canoes. There was a total of 39 trips recorded in this category with an average dive time of 6.8 hrs/trip. The combined total duration for trips of this type was 265.2 hrs.

- Trip type 2: Fishers harvesting lobster and giant clam as the main target species to sell to a local fishing company, with holothurians collected opportunistically. Fishers operating from sailing canoes. There were a total of 37 trips recorded in this category with an average dive time of 10.4 hrs/trip. The combined total duration for trips of this type was 384.1 hrs.

- Trip type 3: Fishers harvesting lobster and giant clam as the main target species, with holothurians collected opportunistically. Using dugout and outrigger canoes, fishers were launched from and picked up by a local fishing vessel. There was a total of 45 trips recorded in this category with an average dive time of 3.9 hrs/trip. The combined total duration for trips of this type was 174.1 hrs.

Major fishing camps were set up at Nabaina Island in the west and Nagobi Island in the east Long/Kosmann Reef area. Enivala Island in the Bramble Haven was the major base camp. Due to declining availability and increasing commercial value of marine resources, the frequency of territorial disputes has increased between different clans and villages over exactly who has the right to fish in this area (see Kinch, 2000). Fishers from other areas are now manipulating clan and kin ties to gain access to these waters where remaining stocks are still to be found, adding pressure to what holothurian stocks are present (Kinch, 2002a).

Harvesting is done either by hand-collection or free diving. A typical dive day starts early in the morning with boats leaving for harvesting areas and outer reefs. Fishers utilize accumulated knowledge and skill about the local territory; especially important are the tides and associated wind-current relationships, which help to define access and availability of species. With favorable weather conditions, clear sky, calm sea, and non-turbid water, holothurians can be collected in water up to 25 m in depth (Kinch, 2002a).

At each fishing site, fishers (this may include women, who either actively harvest or act as spotters) enter the water. If the water is shallow enough, fishers wade across the reef collecting them. If in slightly deeper water, they dive down and lance them with a lengthened spear. If the water is deeper, a small harpoon embedded in a lead weight is dropped. The small harpoon punctures the holothurian, which is then retrieved and brought to the surface.

Boats are typically out on the water for most of the day, but actual dive time averages between 3 and 4 hours. Upon returning to the island base camp, processing of the beche-de-mer commences while the women prepare the evening meal. At least one man watches over the drying racks to ensure that the beche-de-mer is dried and smoked properly.

Dinghies are now beginning to take the place of sailing canoes. The increase in dinghy use represents an increase in household income due to the value of beche-de-mer and the need to range further to locate fresh stocks. It also enables fishers to harvest faster and operate over greater distances. On Brooker Island in late 1999 (Kinch, 1999), there was only one dinghy; in early 2002 there were 22 dinghies in use (Kinch, 2002a).

Catch per Unit Effort (CPUE) values are listed below for individual trip types by day and by species (Table 5.11 and Figure 5.7). Trip type 1 generally had the highest CPUE, not an unexpected result considering that holothurians were the main target species.

During the 2001 CSIRO/NFA/CI stock assessment, the mean density for all commercial species across Milne

Bay Province was 21.24/ha, with high value species having a mean density of 5.22/ha (Skewes *et al.*, 2002; Kinch, 2002a). For the areas of Long/Kosmann Reef and Bramble Haven the mean density for commercial holothurians from this stock assessment were as follows:

- Long/Kosmann Reef west area had a mean density of 18.64/ha on the reef edge (20 sites surveyed) covering an area of 3,421.11 ha; and 22.06/ha on the reef top (51 sites surveyed) covering an area of 10,091.22 ha.

- Long/Kosmann Reef east area had a mean density of 13.54/ha on the reef edge (24 sites surveyed) covering an area of 1,968.64 ha; and 25.74/ha on the reef top (34 sites surveyed) covering an area of 6,888.61 ha.

- Bramble Haven area had a mean density of 6.94/ha on the reef edge (18 sites surveyed) covering an area of 1,036.13 ha; and 12.50/ha on the reef top (21 sites surveyed) covering an area of 4,457.94 ha (Kinch, in prep).

Overall, the mean densities for holothurians are comparable to those of other parts of Milne Bay Province, though this area was noted during the CSIRO/NFA/CI stock assessment to have been affected by heavy fishing pressure, and the CPUE rates reflect this. Also, low-value species now dominate the catch.

Table 5.11. Summary of holothurian harvest by Brooker Islanders for different day trip types, Jan.–May '99. CPUE = Catch Per Unit Effort.*

Species		Trip Type 1		Trip Type 2		Trip Type 3	
Common Name	Scientific Name	No. harvested	CPUE	No. harvested	CPUE	No. harvested	CPUE
Amberfish	*Thelenota anax*	103	0.1	-	-	-	-
Black teatfish	*Holothuria nobilis*	133	0.2	225	0.1	43	0.2
Blackfish	*Actinopyga miliaris*	20	<0.05	10	<0.05	3	<0.05
Brown sandfish	*Bohadschia vitiensis*	2	<0.05	25	<0.05	4	<0.05
Curryfish	*Stichopus variegatus*	97	0.1	-	-	-	-
Elephant trunkfish	*Holothuria fucsopunctata*	148	0.2	-	-	-	-
Greenfish	*Stichopus chloronotus*	164	0.3	47	<0.05	-	-
Lollyfish	*Holothuria atra*	257	0.3	438	<0.02	-	-
Prickly redfish	*Thelenota ananas*	214	0.2	66	<0.05	30	<0.05
Stonefish	*Actinopyga lecanora*	-	-	5	<0.05	-	-
Surf redfish	*Actinopyga mauritiana*	6	<0.05	-	-	-	-
Tigerfish	*Bohadschia argus*	186	0.4	104	<0.05	-	-
White teatfish	*Holothuria fuscogilva*	156	0.2	3	<0.05	16	<0.05

*CPUE = number of individual animals caught per person per hour by trip type.

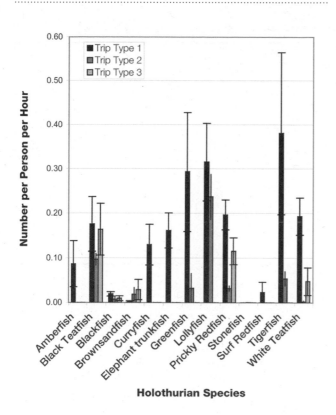

Figure 5.7. CPUE for holothurian species harvested by Brooker Islanders for different day trip types, Jan.–May '99. See Table 5.11 for scientific names.

Giant clams (Tridacna) and other shells

The commercial giant clam fishery developed in Milne Bay Province in the wake of the reduction of illegal fishing by Taiwanese vessels and in response to sustained demand. The Milne Bay Fisheries Authority (MBFA), established in 1979 (Munro, 1989), began the export of giant clams from the province in 1983 (Lokani and Ada, 1998). A ban on the purchase and export of wild-caught giant clam meat was later imposed in May 1988 but lifted in May 1995. The ban allowed for some regeneration of giant clam stocks, thus providing an incentive for a local fishing company to commence harvesting and exports. This ban was reimposed in 2000 when it was found that a local fishing company was infringing on its licensing arrangements (Kinch, 2001b and 2002c).

Fishing methods for giant clams are simple, owing to the shallow distribution, conspicuous appearance, and sedentary habits of these bivalves. Small clams are collected opportunistically during reef-gleaning activities and non target-specific fishing, while larger specimens are collected by free diving. The flesh is excised from the shell by slipping a knife or sharpened wooden stick along its inner surface to cut one end of the adductor muscle. Giant clams located in deeper water are hauled to the surface using ropes and chains. The mantle and muscle are then removed and the shell is dropped back into the sea. Previously, a local fishing company aided village divers in harvesting giant clams by towing canoes to harvesting sites and winching specimens to the surface from their boats.

Giant clams are susceptible to fishing pressure, and once populations are reduced below certain undefined levels, stocks will become non-sustaining. Depending on factors such as prevailing currents and isolation of reefs, the re-establishment of stocks may take hundreds of years (Munro, 1993). Fishing pressure on clam populations in traditional fishing areas was suggested to be at maximum levels by some observers, and this has been confirmed by the CSIRO/NFA/CI stock assessment.

From January to the end of September 1999, a local fishing company purchased 697 kg of giant clam muscles (mostly *Tridacna gigas* and *T. derasa*) from Brooker Islanders. During this period, a local fishing company was purchasing two sizes of giant clam muscle based on weight. Tridacnid specimens weighing less than 400 g fetched a price of 6 kina, and those weighing more than 400 g fetched a price of 10 kina. A total of 551 kg (or 1,970 clams) of clams weighing less than 400 g were sold between January and September 1999, earning 3,306 kina. During the same period, 146 kg (or 170 clams) of clams weighing more than 400 g were sold, earning 1,460 kina (Kinch, 1999). Of this volume, almost a third of the *T. gigas* sold were not full-grown adults.

A more detailed survey from 5 January to 1 May 1999 was conducted by J. Kinch (Kinch, in prep.). During this period 121 trips were recorded where fishers from Brooker Island were targeting giant clams, holothurians, and crayfish in the Long/Kosmann Reef area surrounding Nagobi and Nabaina Islands, and the Bramble Haven Group (see above for details). Catches of other commercial and utilitarian shells were also recorded throughout this period, and the CPUE rates are provided for each species for three different day trip types (Table 5.12 and Figure 5.8).

The most commonly harvested clam during the first part of 1999 were species of the genus *Hippopus*, which made up the bulk of the clams in the unidentified category. These clams are not for commercial sale, but are utilized for subsistence and in trade.

Trochus shells are a commercially valuable resource; the lustrous nacre is used in the manufacture of items such as buttons. Brooker Islanders collect these shells for sale either to trade-store owners in the village, or directly to exporters. The flesh of the trochus shell is also frequently consumed. The black-lip oyster (*Pinctada margaritifera*) is another mollusc collected for commercial and subsistence purposes. The shell Charonia tritonis is collected for sale to passing yachts or as a trumpet once a hole is made near the spiral point.

Shell species that are mostly consumed in households include the commercially harvested species of shells, and noncommercial shells such as cockles, abalone, oysters, *Lambis* spp., *Trochus maculatus*, *Turbo* spp. including *Turbo cinereus*, *Haliotis* spp., *Cypraea caputserpentis*, *Cypraea arabica*, *Cypraea testudinaria*, *Cerithium nodulosum*, and

Table 5.12. Summary of shellfish harvest by Brooker Islanders for different day trip types, Jan.– May '99. CPUE = Catch Per Unit Effort.*

Species	Trip Type 1		Trip Type 2		Trip Type 3	
	No. harvested	CPUE	No. harvested	CPUE	No. harvested	CPUE
Pinctada margaritifera	9	0.018	81	0.042	18	0.078
Clam *H. hippopus*	78	0.071	99	0.092	14	0.037
Clam *T. derasa*	64	0.073	10	0.008	85	0.310
Clam *T. gigas*	9	0.011	2	<0.002	39	0.159
Clam *T. squamosa*	-	0.003	3	<0.002	3	0.007
Clam Unidentified	-	0	781	0.489	4	0.019
Shell *C. tritonis*	-	0	5	0.002	1	0.003
Shell *Lambis* spp.	-	0	2	<0.002	2	0.004
Trochus spp.	14	0.043	169	0.148	5	0.014

*CPUE = number of individual animals caught per person per hour by trip type.

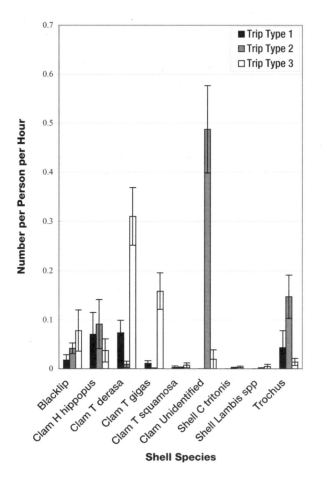

Figure 5.8. CPUE for shellfish species harvested by Brooker Islanders for different day trip types, Jan.–May '99.

Strombus luhuanus (Kinch, 1999). Of these, *S. luhuanus* is a major subsistence food. Shells of this species are processed and threaded onto stems of coconut fronds and are a predominant trade item, with hundreds or thousands collected per session. Women and girls are the predominant collectors, but men help occasionally.

Fishes

Brooker Islanders utilize fishes, both as a food source and as a commercial resource for income. Of the 7,786.4 kg of fishes caught from July 1998–September 1999, 5,703.6 kg (or roughly 75%) were purchased by buyers on the island of Misima, at a total value of just over 10,000 kina. Fishing techniques employed by the locals include trolling, hand lining, netting, and spearing. Most of the annual fish catch comes from net fishing (Figure 5.9).

A wide variety of species are caught for consumption on Brooker Island, and of these, the ox-eye scad (*Selar boops*) is the most commonly consumed. Other frequently consumed fishes are listed below (Table 5.13).

Fish consumption is seasonally dependent. For example, in the months of June and July, the silver spinefoot or "vivial" (*Siganus* sp.) makes up a greater portion of the fish consumption as it is abundantly netted at this time. Also, fish consumption is proportionally higher among groups who are camping on islands during expeditions than in the village.

It should be noted that some species (e.g., *Selar boops*) that contribute heavily to the diet were not recorded in the coral reef fisheries survey during the RAP. This is due to habitat preferences and behavioral patterns exhibited by such species. They do not commonly frequent coral reef habitats and are mainly nocturnally active.

Table 5.13. Most commonly consumed fishes on Brooker Island (source: J. Kinch PhD field notes).

Misima name	Scientific name	Misima name	Scientific name
Atuni	*Selar boops*	Kitun	Siganidae (rabbitfishes)
Tupatupa	Carangidae (trevallies)	Anuwal/Kanivala	Sphyraenidae (barracudas)
Waloya/Pilihul	Mugilidae (mullets)	Kingfish	*Seriola* spp.
Tuna	Scombridae (tunas)	Tokeli	Lethrinidae (emperors)
Yalyal	*Grammatorcynos bilineatus*	Kabela	*Acanthurus auranticavus*
Yesimoli	*Naso tuberosus*	Ahiat	*Lethrinus erythracanthus*
Vivilal	*Siganus* sp.	Yabwau	*Lutjanus gibbus*
Buhmanawi	*Lethrinus olivaceus*	Mwananuya	Serranidae (groupers/cods)
Suwa	*Elegatis bipinnulatus*	Labeta	*Lethrinus nebulosus*
Malawi	*Acanthurus dussumieri*		

Crayfish

The crayfish species currently being harvested include the double-spined ornate lobster (*Panulirus penicillatus*), the spiny lobster (*P. ornatus*), the painted coral lobster (*P. versicolor*), and the long-legged spiny lobster (*P. longipes*). The first two species are the most commonly exploited. Brooker Islanders capture crayfish by spearing them while free diving on the reef slope and crest. Collections are sometimes made at night with the aid of underwater torches (flashlights).

A total of 7,105 crayfish were harvested by Brooker Islanders during the recorded period of January 1998–September 1999. A local fishing company purchased the entire amount at a total value of 11,372 kina.

From 5 January to 1 May 1999, 121 trips were recorded where fishers from Brooker Island were targeting giant clams, holothurians, and crayfish in the Long/Kosmann Reef area surrounding Nagobi and Nabaina Islands, and the Bramble Haven Group (see above for details). The CPUE for trip type 1 was 0/hr (5 crayfish caught); trip type 2 was 0.3/hr (661 crayfish caught); and trip type 3 was 2.0/hr (407 crayfish caught) (Kinch, in prep.).

Commercial value of marine resources

The importance of marine resources to communities in Milne Bay province cannot be overstated. Between the months of January and September 1999, a total of 56,649 kina was earned through the sale of marine resources. Of the various types of marine organisms that are commercially exploited, the beche-de-mer fishery is by far the most important. The sale of trochus, crayfish, fishes, and clams also contribute significantly to the income on Brooker (Table 5.14).

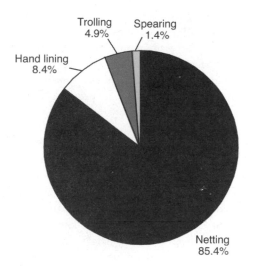

Figure 5.9. Contribution of different fishing techniques to overall fish catch weight on Brooker Island (Sept.'98–Oct.'99). Source: Kinch, 1999.

CONCLUSIONS

Conservation of biological marine resources is an issue that, until recently, has not been of major concern to remote island and coastal communities across PNG. Given the vastness of the marine territory, coupled with low human population densities, this is understandable. However, population growth and rising resource value has changed the situation to the point where conservation and sustainable resource management is now a major concern.

Remote communities, such as that on Brooker Island, are now exploiting significant quantities of marine resources, often in ways that seem unsustainable. For example, there

Table 5.14. Income (rounded to nearest kina) on Brooker Island from the sale of marine resources, January–September 1999 (source: J. Kinch field notes and commercial buyers).

Month	Fishes	Beche-de-mer	Trochus	Blacklip	Shark-fin	Crayfish	Clam >400 g	Clam <400 g	Total
Jan.	-	639	2,437	56	52	2,364	411	65	6,024
Feb.	-	7,277	2,526	210	-	-	-	-	10,013
March	21	834	-	-	-	2,920	681	310	4,766
April	55	9,735	1,013	-	151	2,252	1,983	925	16,114
May	859	5,106	499	71	-	20	18	45	6,618
June	4,009	196	-	-	-	256	6	20	4,487
July	1,539	2,109	707	162	113	356	204	95	5,285
August	195	554	890	-	-	-	-	-	1,639
Sept.	1,192	14	382	-	-	112	3	-	1,703
Total	**7,870**	**26,464**	**8,454**	**499**	**316**	**8,280**	**3,306**	**1,460**	**56,649**

is a tendency to over-fish certain species simply because a local buyer is available on site or market prices are high. This represents a potentially serious risk to the species that are harvested. Sedentary species such as clams, trochus, and holothurians are particularly threatened with over-exploitation. A community survey of Brooker peoples' perception of the status of their marine resources indicated that several traditional fishing areas are generally believed to be depleted in holothurians, trochus, blacklip, and clams. Even more alarming is the fact that most households admitted at the time of the community survey to not following or being aware of government restrictions on resource harvesting.

Strict policing of government-imposed restrictions is an impossibility given the remoteness of communities. Community workshops aimed at increasing awareness of resource management issues among the artisanal sector have taken place recently, and these appear to be the most effective solution. However, even though there is growing awareness of resource depletion and management needs among these communities, this does not necessarily equate to implementation. Similar to the situation in western societies, the requirement to earn money is a force that often outweighs the desire for conservation and proper management of resources. Conservation International is now counteracting this through it's Community Engagement Program which generates awareness of resource decline through education, and assists communities in making management plans for sustainable harvesting and the development of Locally Managed Marine Areas (LMMAs).

It is encouraging that the recent community survey on Brooker revealed an interest in the idea of rotating access to fishing areas as a means of conserving resources (see Kinch, 2001a). Throughout Milne Bay Province the practice of closing reefs or fishing grounds has been carried out for a certain length of time following a death. After a period

of several months to several years the area is reopened and people can access that area for harvesting purposes again. People are thus well aware of the benefits of such reef closure in resource regeneration, and this practice therefore possibly offers the most culturally appropriate way to introduce community managed Marine Protected Areas (MPAs). This practice can also be applied for other reasons, as a recent example from Skelton Island in the Engineer Group shows. The village there had decided to apply this traditional practice of closure to allow the numbers of beche-de-mer and trochus to replenish so money could be made available for a new church building (Kinch, 2002e).

The current study and other programs being undertaken by Conservation International in Milne Bay Province are an important first step in the conservation and management of the region's biodiversity. Further biological study and monitoring of marine resources is needed to ensure that exploitation by the artisanal and commercial fishing sectors occurs in a sustainable manner. Furhter detailed socio-cultural studies (such as that understaken by J. Kinch at Brooker Island) will assist tremendously in a proper understanding of the management status of artisanal fisheries.

REFERENCES

Allen, G. 1997. Marine Fishes of Tropical Australia and South-East Asia. Perth: Western Australian Museum.

Allen, G. R. and Steene, R. C. 1994. Indo-Pacific Coral Reef Field Guide. Singapore: Tropical Reef Research.

Allen, G. R. and R. Swainston. 1992. Reef Fishes of New Guinea. Madang, Papua New Guinea: Christensen Research Institute.

Australian Marine Science and Technology. 1997. Papua New Guinea, Western and Gulf Provinces Coastal Zone

Management Plan: feasibility study completion report. Canberra: Australian Agency for International Development.

Chesher, R. H. 1980. Stock Assessment: Commercial Invertebrates of Milne Bay Coral Reefs. Unpublished report for Fisheries Division, Department of Primary Industries, Papua New Guinea.

Colin, P. L. and C. Arneson. 1995. Tropical Pacific Invertebrates. Beverly Hills, California: Coral Reef Press.

Conand, C. 1998. Holothurians. In: Carpenter, K.E. and V.H. Niem (eds.). FAO species identification guide for fishery purposes. The living marine resources of the Western Central Pacific. Volume 2. Cephalopods, crustaceans, holothurians and sharks. Rome, FAO. Pp. 687–1396.

CSIRO. 2001. Research for Sustainable Use of Beche-de-mer Resources in Milne Bay Province, Papua New Guinea. Project Document, FIS/2001/059 submitted to ACIAR.

Dartnall, H. J. and M. Jones. 1986. A manual of survey methods of living resources in coastal areas. ASEAN-Australia Cooperative Programme on Marine Science Hand Book. Townsville: Australian Institute of Marine Science.

Gosliner, T. M., D. W. Behrens, and G. C. Williams. 1996. Coral Reef Animals of the Indo-Pacific. Monterey: Sea Challengers.

Ingles, J. 2001. Fisheries of the Calamianes Islands, Palawan Province, Philippines. In: Werner, T.B. and G.R. Allen (eds.). A Rapid Marine Biodiversity Assessment of the Calamianes Islands, Palawan Province, Philippines. Bulletin of the Rapid Assessment Program 17. Washington, DC: Conservation International.

Kinch, J. In prep. Fishing Efforts in the Long-Kosmann Reef and Bramble Haven Areas of Milne Bay Province, Papua New Guinea: A Case Study of Brooker Island Fishers. Unpublished Manuscript.

Kinch, J. 2002a. An Overview of the Beche-de-mer Fishery in Milne Bay Province, Papua New Guinea. SPC Beche-de-mer Bulletin. 17: 2–16.

Kinch, J. 2002b. Information Pamphlets: Beche-de-mer; Clam; Trochus; Mud Crabs; Lobster; Sharks; Turtles; Mangroves; Seagrasses; Oil Pollution; Rubbish Disposal; and Fishing Zones and Limits. Unpublished educational materials prepared for Conservation International, Alotau, Milne Bay Province, Papua New Guinea.

Kinch, J. 2002c. Giant Clams: Their Status and Trade in Milne Bay Province, Papua New Guinea. Traffic Bulletin. 19 (2): 67–75.

Kinch, J. 2002d. The Development of a Monitoring Program for the Management and Sustainable Use of Sea Turtle Resources in Milne Bay Province, Papua New Guinea. A Proposal prepared for the South Pacific Regional Environment Program, Apia, Western Samoa. Pp: 39.

Kinch, J. 2002e. The Role of Traditional Reef Closure for Conservation, Milne Bay, Papua New Guinea. In: WWF: Mainstreaming Nature Conservation in the South Pacific: The NGO Experience. Pp: 11–12. Suva: World Wild Fund for Nature-South Pacific.

Kinch, J. 2001a. Social Feasibility Study for the Milne Bay Community-Based Coastal and Marine Conservation Program. Unpublished report to the United Nations Milne Bay Community-Based Coastal and Marine Conservation Program, PNG/99/G41, Port Moresby, Papua New Guinea.

Kinch, J. 2001b. Clam Harvesting, the Convention on the International Trade in Endangered Species (CITES) and Conservation in Milne Bay Province. SPC Fisheries Newsletter 99:24–36.

Kinch, J. 2000. Brooker Island Versus Ware Island: A Report on the Ongoing Dispute over the Nabaina and Nagobi Islands and the Long/Kosmann Reefs; Milne Bay Province. Unpublished Report to the Administrator's Office, Milne Bay Provincial Government, Alotau, Milne Bay Province, Papua New Guinea.

Kinch, J. 1999. Economics and Environment in Island Melanesia: A General Overview of Resource Use and Livelihoods on Brooker Island in the Calvados Chain of the Louisiade Archipelago, Milne Bay Province, Papua New Guinea. Unpublished report prepared for Conservation International, Port Moresby, Papua New Guinea.

Kinch, J., D. Mitchell, and P. Seeto. 2001. Information Paper for Milne Bay Province Wide Stock Assessment and Biogeographical Survey. Unpublished report prepared for Conservation International, Washington DC.

La Tanda. In press. Coral reef fish stock assessment in the Togean and Banggai Islands, Sulawesi, Indonesia. In: G.R. Allen and S. A. McKenna, (eds.). A Rapid Biodiversity Assessment of the Coral Reefs of the Togean and Banggai Islands, Sulawesi, Indonesia. RAP Bulletin of Biological Assessment. Washington, DC: Conservation International.

Lanyon, J., C. Limpus, and H. Marsh. 1989. Dugongs and Turtles: Grazers in the Seagrass System. In: A. Larkum, A. McComb, and S. Shepherd (eds.). Biology of Seagrasses. Amsterdam: Elsevier. Pp. 610–634.

Ledua, E., S. Matoto, P. Lokani, and L. Pomat. 1996. Giant Clam Resource Assessment in Milne Bay Province. Unpublished report for South Pacific Commission and the National Fisheries Authority (Papua New Guinea).

Lieske, E. and R. Myers. 1994. Coral Reef Fishes – Indo-Pacific & Caribbean. London: Harper Collins.

Lokani, P. and K. Ada. 1998. Milne Bay Province: Product Exports – 1997. Unpublished report for National Fisheries Authority (Papua New Guinea).

Munro, J. 1989. Development of a Giant Clam Management Strategy for Milne Bay Province. Unpublished report for Department of Fisheries and Marine Resources, Papua New Guinea.

Omeri, N. 1991. Fisheries and Marine Policy for Milne Bay Province. Unpublished report for Department of Fisheries and Marine Resources, Papua New Guinea.

Randall, J. E., G. R. Allen, and R. C. Steene. 1990. Fishes of the Great Barrier Reef and Coral Sea. Bathurst, Australia: Crawford House Press.

Skewes, T., Kinch, J., Polon, P., Dennis, D., Seeto, P., Taranto, T., Lokani, P., Wassenberg, T., Koutsoukos, A, Sarke, J. 2002. Research for the Sustainable Use of Beche-de-mer Resources in Milne Bay Province, Papua New Guinea. CSIRO Division of Marine Research Final Report, Cleveland Australia.

Sparre, P. and S. Venema. 1992. Introduction to tropical fish stock assessment. Part 1. FAO Fisheries Technical Paper 306. Food and Agricultural Organization of the United Nations, Rome.

Werner, T. B. and G. R. Allen (eds.). 1998. A rapid biodiversity assessment of the coral reefs of Milne Bay Province, Papua New Guinea. RAP Working Papers 11, Washington, DC: Conservation International.

Appendices

Appendix 1

List of coral species recorded at Milne Bay Province, Papua New Guinea during 2000 RAP survey

Douglas Fenner and Emre Turak

Sites 1–28 include observations of both D. Fenner and E. Turak.
Summary of top coral diversity sites are presented in the map following the table.

Species	Site records
Family Astrocoeniidae	
Stylocoeniella armata (Ehrenberg, 1834)	4, 5, 9, 12, 13, 14, 16, 18, 19, 20, 22, 29, 30, 31, 32, 35, 37, 39, 43, 44, 45, 46, 48, 49, 50, 51, 52, 55, 57
Stylocoeniella guentheri (Basset-Smith, 1890)	1, 6, 8, 9, 11, 13, 15, 18, 22, 23, 27, 28, 31, 33, 36, 37, 39, 43, 46, 48, 52, 53, 54, 56, 57
Family Pocilloporidae	
Madracis kirbyi Veron & Pichon, 1976	27
Palauastrea ramosa Yabe and Sugiyama, 1941	2, 12, 13, 20, 21, 22, 23, 37, 38, 39, 50, 51
Pocillopora ankeli Scheer and Pillai, 1974	8, 15
Pocillopora damicornis (Linnaeus, 1758)	1, 2, 3, 4, 5, 6, 7, 8, 9, 11, 12, 13, 14, 15, 16, 17, 18, 19, 20, 21, 22, 23, 24, 25, 26, 27, 28, 29, 32, 33, 34, 35, 37, 38, 39, 40, 41, 42, 43, 45, 46, 48, 50, 51, 54, 55, 56, 57
Pocillopora danae Verrill, 1864	5, 27, 29, 31, 32, 33, 35, 38, 48, 49, 52, 53
Pocillopora elegans Dana, 1846	37
Pocillopora eydouxi Milne Edwards and Haime, 1860	1, 4, 5, 6, 8, 9, 16, 17, 18, 19, 22, 25, 27, 28, 29, 33, 34, 39, 43, 44, 47, 49, 50, 52, 53, 54, 55, 56
Pocillopora meandrina Dana, 1846	1, 2, 4, 6, 9, 12, 14, 15, 17, 18, 19, 22, 24, 25, 28, 29, 30, 32, 39, 43, 44, 45, 47, 48, 49, 50, 52, 53
Pocillopora verrucosa Ellis & Solander, 1786	1, 2, 3, 4, 5, 7, 8, 9, 11, 12, 14, 15, 16, 17, 18, 19, 20, 21, 22, 23, 24, 25, 26, 27, 28, 29, 30, 32, 33, 34, 35, 37, 39, 40, 41, 42, 43, 44, 45, 46, 47, 48, 49, 52, 53, 54, 55, 56
Pocillopora woodjonesi Vaughan, 1918	6, 18
Seriatopora aculeata Quelch, 1886	1, 4, 5, 6, 7, 12, 30, 39, 53
Seriatopora caliendrum Ehrenberg, 1834	1, 2, 3, 7, 8, 10, 11, 12, 13, 14, 15, 16, 17, 20, 21, 22, 24, 25, 27, 32, 41, 48, 57
Seriatopora hystrix Dana, 1864	1, 2, 3, 4, 5, 6, 7, 8, 9, 10, 11, 12, 13, 14, 15, 16, 17, 18, 19, 20, 21, 22, 23, 24, 25, 26, 27, 28, 29, 30, 31, 32, 33, 34, 35, 37, 38, 39, 40, 42, 43, 44, 45, 46, 48, 49, 50, 51, 52, 54, 55, 56, 57
Seriatopora dendritica Veron, 2000	1, 3, 4, 5, 11, 33, 51
Seriatopora sp. 1	

continued

Species	Site records
Stylophora subseriata (Ehrenberg, 1834)	1, 2, 3, 4, 5, 6, 7, 8, 9, 10, 11, 12, 13, 14, 15, 16, 17, 18, 19, 20, 21, 22, 23, 24, 25, 26, 27, 28, 29, 30, 31, 32, 33, 34, 35, 36, 37, 38, 39, 40, 41, 43, 44, 45, 46, 48, 49, 50, 51, 53, 54, 55, 56, 57
Stylophora pistillata (Esper, 1797)	5, 6, 9, 10, 14, 16, 17, 18, 29, 43, 44, 47, 48, 52, 53, 54
Stylophora sp. 1	4, 5, 6, 9, 15, 17, 18
Family Acroporidae	
Acropora abrolhosensis Veron 1985	11
Acropora abrotanoides Lamarck, 1816	8, 16, 17, 18, 43, 44, 48
Acropora aculeus (Dana, 1846)	1, 2, 3, 4, 5, 6, 7, 8, 9, 10, 11, 12, 13, 14, 15, 16, 17, 18, 19, 20, 22, 24, 25, 26, 28, 29, 30, 32, 33, 34, 35, 36, 37, 38, 39, 40, 42, 43, 45, 46, 48, 49, 50, 51, 52, 54, 55, 56, 57
Acropora acuminata (Verrill, 1864)	1, 3, 4, 10, 14
Acropora anthocercis (Brook, 1893)	4, 5, 19, 27
Acropora aspera (Dana, 1846)	15
Acropora austera (Dana, 1846)	3, 11, 16, 17, 18, 19, 20, 21, 22, 23, 24, 25, 26, 27, 28, 43, 48, 54, 56
Acropora cf. *awi* Wallace and Wolstenholme, 1998	18, 19
Acropora batunai Wallace, 1997	26, 28
Acropora bruggemanni (Brook, 1893)	30, 54
Acropora carduus (Dana, 1846)	3, 6, 8, 11, 13, 14, 16, 18, 25, 26, 27, 28, 40, 48, 54, 56
Acropora caroliniana Nemenzo, 1976	2, 4, 5, 12, 14, 16, 20, 22, 26, 28, 29, 32, 34, 35, 37, 38, 39, 40, 42, 43, 46, 51, 54, 56
Acropora cerealis (Dana, 1846)	1, 2, 4, 5, 6, 7, 8, 9, 10, 11, 12, 14, 15, 16, 17, 18, 20, 21, 22, 23, 24, 25, 26, 27, 28
Acropora clathrata (Brook, 1891)	4, 5, 6, 8, 9, 15, 16, 18, 19, 20, 29, 30, 32, 33, 37, 39, 43, 47, 48, 49, 53, 54, 55, 56
Acropora cylindrica Veron & Fenner, 2000	2, 7, 8, 10, 11, 12, 13, 14, 15, 16, 17, 18, 20, 22, 23, 24, 25, 26, 27, 28, 33, 35, 36, 38, 41, 46, 49, 50, 56
Acropora crateriformis (Gardiner, 1898)	8
Acropora cuneata (Dana, 1846)	4, 12, 15, 17, 21, 25, 36, 48, 55
Acropora cytherea (Dana, 1846)	4, 5, 6, 8, 9, 16, 18, 25, 32, 35, 37, 38, 39, 40, 42, 43, 48
Acropora dendrum (Bassett-Smith, 1890)	34, 36, 46, 56
Acropora cf. *desalwii* Wallace, 1994	28
Acropora digitifera (Dana, 1846)	8, 9, 11, 12, 14, 15, 17, 18, 22, 25, 27, 28, 29, 36, 39, 41, 43, 44, 47, 49, 50, 52, 53, 54, 55, 57
Acropora divaricata (Dana, 1846)	2, 4, 5, 6, 7, 8, 11, 14, 16, 18, 19, 20, 22, 24, 39, 43, 55
Acropora echinata (Dana, 1846)	3, 6, 7, 8, 13, 16, 17, 18, 19, 20, 22, 23, 24, 26, 27, 39, 43, 55
Acropora elegans Milne Edwards and Haime, 1860	2, 25
Acropora elseyi (Brook, 1892)	3, 7, 8, 14, 15, 17, 18, 20, 21, 22, 24, 27, 28, 50, 54, 56
Acropora florida (Dana, 1846)	3, 4, 5, 6, 8, 9, 10, 12, 13, 14, 15, 16, 17, 18, 19, 20, 21, 22, 23, 25, 28, 29, 31, 32, 34, 36, 39, 43, 47, 48, 54, 55, 56
Acropora formosa (Dana, 1846)	1, 2, 3, 4, 6, 8, 9, 10, 12, 13, 14, 16, 17, 18, 19, 20, 21, 22, 23, 24, 25, 26, 27, 28, 29, 31, 32, 33, 34, 35, 36, 38, 40, 42, 44, 46, 48, 49, 50, 56, 57
Acropora gemmifera (Brook, 1892)	4, 5, 6, 7, 11, 12, 14, 15, 16, 17, 18, 19, 20, 21, 22, 24, 25, 26, 29, 33, 36, 39, 43, 45, 47, 48, 49, 52, 54, 55, 57
Acropora cf. *glauca* (Brook, 1893)	19

continued

Species	Site records
Acropora grandis (Brook, 1892)	3, 8, 9, 10, 11, 12, 13, 16, 17, 19, 20, 21, 22, 23, 25, 26, 27, 28, 32, 33, 38, 39, 40, 42, 46, 48, 51, 54, 56, 57
Acropora granulosa (Milne Edwards and Haime, 1860)	1, 2, 3, 4, 5, 6, 7, 8, 9, 10, 11, 12, 13, 14, 15, 16, 17, 18, 20, 22, 23, 24, 25, 26, 27, 28, , 29, 32, 33, 34, 35, 37, 43, 49, 54
Acropora horrida (Dana, 1846)	3, 5, 16, 18, 20, 34, 39, 54, 56
Acropora humilis (Dana, 1846)	1, 2, 3, 4, 5, 6, 7, 8, 9, 10, 11, 12, 14, 15, 16, 17, 18, 19, 20, 21, 23, 24, 25, 26, 27, 28, 29, 30, 31, 32, 33, 34, 35, 36, 37, 38, 39, 40, 42, 43, 44, 45, 46, 47, 48, 49, 50, 51, 53, 54, 55, 57
Acropora hyacinthus (Dana, 1846)	4, 15-19, 22-23, 25-29, 31, 34-40, 42-44, 46, 49-50, 52, 54-57
Acropora jacquelineae Wallace, 1994	1, 3, 14, 16, 20, 22, 23, 24, 27, 28, 38
Acropora latistella (Brook, 1892)	5, 6, 16, 29, 30, 32, 33, 34, 35, 36, 39, 43, 44, 45, 47, 49, 50, 52, 55, 57
Acropora listeri (Brook, 1893)	14
Acropora lokani Wallace, 1994	7, 15, 17, 20, 22, 25, 32, 38, 39, 40, 42, 46, 51, 57
Acropora longicyathus (Milne Edwards and Haime, 1860)	3, 4, 6, 7, 8, 10, 11, 12, 13, 15, 16, 17, 20, 22, 24, 25, 26, 27, 28, 35, 36, 38, 40, 48, 51, 54, 56, 57
Acropora loripes (Brook, 1892)	1, 2, 3, 4, 5, 6, 7, 8, 9, 10, 12, 14, 15, 16, 17, 18, 19, 21, 22, 23, 25, 26, 27, 28, 29, 30, 33, 34, 35, 39, 40, 43, 46, 48, 50, 54, 57
Acropora lutkeni Crossland, 1952	4, 8, 9, 10, 12, 14, 16, 17, 18
Acropora microclados (Ehrenberg, 1834)	4, 5, 6, 9, 14, 16, 18, 19, 20, 27, 28, 29, 38, 39
Acropora microphthalma (Verrill, 1859)	4, 5, 8, 11, 14, 15, 16
Acropora millepora (Ehrenberg, 1834)	2, 3, 4, 7, 8, 9, 11, 12, 13, 14, 15, 16, 17, 18, 19, 20, 21, 22, 23, 24, 25, 26, 27, 28, 29, 32, 33, 34, 35, 36, 37, 38, 39, 40, 41, 42, 43, 44, 45, 46, 47, 49, 50, 51, 52, 53, 54, 55, 56, 57
Acropora monticulosa (Brüggemann, 1879)	7, 8, 15, 17, 18, 25, 43, 44, 48, 49, 52, 53
Acropora multiacuta Nemenzo, 1967	10, 11, 21, 29, 31, 32, 33, 34, 40, 42, 45, 46, 57
Acropora nana (Studer, 1878)	8, 18, 21, 44, 52, 53
Acropora nasuta (Dana, 1846)	3, 4, 6, 8, 9, 10, 11, 12, 14, 15, 16, 17, 18, 19, 20, 22, 24, 25, 27, 28, 29, 30, 36, 37, 38, 39, 40, 42, 43, 44, 45, 47, 49, 50, 53, 54, 57
Acropora nobilis (Dana, 1846)	6, 8, 9, 10, 17, 18, 42, 47
Acropora palifera (Lamarck, 1816)	1, 3, 4, 5, 7, 8, 10, 11, 12, 13, 14, 15, 17, 18, 19, 20, 21, 22, 23, 24, 25, 26, 27, 28, 29, 31, 32, 33, 34, 35, 37, 38, 39, 42, 43, 45, 46, 47, 48, 49, 50, 51, 53, 54, 55, 56, 57
Acropora paniculata Verrill, 1902	1, 2, 3, 4, 5, 8, 9, 11, 12, 16, 18, 23, 24, 28
Acropora pichoni Wallace, 1999	41, 42, 57
Acropora plumosa Wallace and Wolstenholme, 1998	1, 2, 3, 4, 5, 6, 7, 8, 10, 11, 12, 13, 14, 15, 16, 17, 18, 20, 22, 23, 24, 25, 27, 28, 29, 32, 34, 35, 38, 39, 40, 42, 46, 48, 56
Acropora pulchra (Brook, 1891)	7, 8, 11, 12, 15, 21, 36
Acropora rambleri (Bassett-Smith, 1890)	
Acropora robusta (Dana, 1846)	8, 12, 14, 17, 18, 19, 25, 27, 29, 43, 44, 45, 47, 48, 49, 52
Acropora rosaria (Dana, 1846)	21, 22
Acropora samoensis (Brook, 1891)	14, 20, 32, 33, 34, 35, 36, 37, 55, 56, 57
Acropora sarmentosa (Brook, 1892)	3, 7, 11, 12, 14, 15, 16, 17, 18, 19, 20, 21, 22, 23, 26, 28, 29, 30, 31, 32, 33, 35, 36, 37, 38, 39, 40, 41, 42, 43, 45, 46, 47, 48, 50, 54, 55, 57
Acropora secale (Studer, 1878)	3, 4, 5, 6, 8, 9, 12, 14, 15, 16, 17, 18, 19, 20, 25, 26, 27, 28, 29, 30, 32, 33, 37, 38, 39, 43, 44, 45, 48, 49, 52, 53, 55

continued

Species	Site records
Acropora sekiseiensis Veron, 1990	24
Acropora selago (Studer, 1878)	2, 3, 4, 5, 6, 7, 9, 10, 12, 13, 14, 16, 17, 18, 20, 22, 24, 25, 29, 32, 34, 35, 36, 38, 42, 46, 51, 54, 56, 57
Acropora seriata Ehrenberg, 1834	17
Acropora spathulata (Brook, 1891)	7, 26, 27, 29, 35, 36, 39, 42, 47, 48, 50, 54
Acropora speciosa (Quelch, 1886)	9, 11, 14, 15, 16, 18, 19, 20, 22, 23, 24, 25, 26, 27, 28
Acropora spicifera (Dana, 1846)	18, 19
Acropora striata (Verrill, 1866)	16, 18, 19
Acropora subglabra (Brook, 1891)	2, 16, 17, 22, 25, 26, 27, 33, 56, 57
Acropora subulata (Dana, 1846)	3, 8, 9, 12, 13, 14, 18, 19, 23, 24, 27, 28
Acropora tenella (Brook, 1892)	41, 57
Acropora tenuis (Dana, 1846)	2, 3, 4, 5, 6, 7, 8, 10, 11, 12, 14, 15, 16, 17, 18, 19, 20, 21, 22, 23, 24, 25, 26, 27, 28, 29, 30, 33, 34, 35, 36, 39, 41, 43, 44, 45, 46, 48, 49, 50, 53, 54, 55, 57
Acropora valenciennesi (Milne Edwards & Haime, 1860)	3, 5, 6, 16, 18, 19, 20, 28, 31, 32, 34, 37, 38, 39, 43, 48, 54, 56, 57
Acropora valida (Dana, 1846)	5, 6, 7, 8, 9, 12, 14, 15, 16, 17, 18, 19, 22, 23, 24, 28, 31, 44, 47, 48, 49, 52, 55
Acropora vaughani Wells, 1954	3, 17, 28, 29, 48, 53, 54, 56
Acropora verweyi Veron and Wallace, 1984	7, 11
Acropora yongei Veron and Wallace, 1984	11, 16, 17, 19, 25, 53
Acropora sp.1	1, 7, 9, 10, 11, 12, 13, 16, 18, 19, 20, 23, 24, 25, 26
Anacropora forbesi Ridley, 1884	2, 7, 11, 20, 22, 28, 56
Anacropora matthai Pillai, 1973	2, 12, 20, 22, 26, 43, 56, 57
Anacropora puertogalerae Nemenzo, 1964	1, 10
Anacropora reticulata Veron and Wallace, 1984	1, 10, 11, 14, 15, 16, 24, 25, 26, 46, 56, 57
Anacropora sp. 1	14, 16, 25
Astreopora cuculata Lamberts, 1980	3, 5, 11, 15, 18, 19, 20
Astreopora expansa Brüggemann, 1877	7, 11, 15
Astreopora gracilis Bernard, 1896	1, 2, 3, 4, 5, 6, 7, 8, 10, 12, 13, 14, 15, 16, 17, 19, 20, 21, 22, 23, 26, 29, 55
Astreopora incrustans Bernard, 1896	21, 23, 24
Astreopora listeri Bernard, 1896	1, 3, 4, 9, 10, 11, 13, 14, 20, 21, 22, 23, 26, 34, 37, 38, 39, 45, 51
Astreopora myriophthalma (Lamarck, 1816)	1, 3, 4, 5, 6, 7, 8, 10, 11, 12, 13, 14, 15, 16, 17, 18, 19, 20, 21, 22, 23, 24, 26, 27, 28, 29, 30, 31, 32, 33, 34, 35, 36, 37, 38, 39, 43, 44, 45, 46, 47, 48, 49, 50, 51, 54, 55, 57
Astreopora ocellata Bernard, 1896	11
Astreopora randalli Lamberts, 1980	2, 6, 7, 10, 11, 12, 13, 14, 15, 16, 18, 19, 20, 21, 23, 24, 25, 27, 29, 31, 32, 33, 34, 35, 36, 37, 38, 39, 40, 41, 43, 44, 45, 46, 49, 50, 51, 53, 54, 57
Astreopora suggesta Wells, 1954	1, 4, 5, 7, 11, 12, 14, 15, 17, 18, 20, 21, 22, 23, 24, 25, 26, 27, 31, 32, 34, 37, 38, 39, 42, 50, 53
Montipora aequituberculata Bernard, 1897	6, 7, 8, 9, 10, 11, 12, 13, 14, 15, 16, 17, 18, 19, 22, 23, 24, 25, 26, 28, 48
Montipora angulata (Lamarck, 1816)	8
Montipora cactus Bernard, 1897	11, 12, 21, 26
Montipora caliculata (Dana, 1846)	10, 15, 17, 20

continued

Species	Site records
Montipora capitata Dana, 1846	8, 11, 12, 18, 19, 22, 26, 28, 37, 45, 48, 49
Montipora capricornis Veron, 1985	2
Montipora cebuensis Nemenzo, 1976	1, 4, 6, 7, 8, 12, 15, 20, 24, 25, 26, 27, 28, 29, 33, 48
Montipora confusa Nemenzo, 1967	4, 5, 6, 13, 18, 22, 23, 25, 27, 28, 32, 37, 38, 43, 48, 49, 54, 55
Montipora corbettensis Veron and Wallace, 1984	19, 43, 48, 55
Montipora crassituberculata Bernard, 1897	22
Montipora danae Milne Edwards & Haime, 1851	7, 10, 12, 14, 19, 20, 22, 27
Montipora delicatula Veron, 2000	7, 13, 14, 17, 18, 19, 24, 25, 26, 27, 57
Montipora efflorescens *Bernard, 1897*	6, 8, 9, 10, 12, 15, 16, 17, 18, 21, 23, 25
Montipora florida Nemenzo, 1967	7, 8, 13
Montipora floweri Wells, 1954	17
Montipora foliosa (Pallas, 1766)	1, 2, 7, 8, 9, 11, 13, 15, 20, 28, 32, 55, 56
Montipora foveolata (Dana, 1846)	4, 8, 11, 12, 14, 17, 18, 21, 22, 31, 35, 43, 44, 45, 47, 48, 49, 51, 54
Montipora grisea Bernard, 1897	1, 4, 6, 9, 11, 12, 14, 15, 18, 20, 22, 24, 25
Montipora hispida (Dana, 1846)	1, 6, 10, 11, 12, 13, 14, 16, 17, 19, 25, 26, 27, 2811, 29, 36, 38, 44, 48, 56
Montipora hodgsoni Veron, 2000	2, 8, 27
Montipora hoffmeisteri Wells, 1954	1, 2, 3, 4, 6, 9, 10, 11, 13, 19, 25, 27
Montipora incrassata (Dana, 1846)	2, 11
Montipora informis Bernard, 1897	1, 2, 3, 4, 5, 6, 7, 8, 10, 11, 12, 13, 16, 17, 18, 20, 21, 22, 23, 24, 25, 26, 28, 29, 32, 33, 35, 36, 51
Montipora mactanensis Nemenzo, 1979	1, 6, 7, 8, 28
Montipora millepora Crossland, 1952	2, 4, 5, 11, 17, 19, 21, 22, 27
Montipora mollis Bernard, 1897	1, 14, 15, 16, 18, 31, 48
Montipora monasteriata (Forsskål, 1775)	3, 5, 7, 9, 12, 14, 15, 16, 18, 19, 20, 22, 24, 25, 28
Montipora niugini Veron, 2000	8, 27
Montipora nodosa (Dana, 1846)	1
Montipora palawanensis Veron, 2000	14, 18, 42, 54
Montipora peltiformis Bernard, 1897	8, 35
Montipora spongodes Bernard, 1897	27
Montipora spumosa (Lamarck, 1816)	11, 14, 21, 51
Montipora stellata Bernard, 1897	11, 26, 27, 38, 57
Montipora tuberculosa (Lamarck, 1816)	1, 3, 6, 7, 8, 11, 12, 16, 18, 19, 20, 21, 22, 23, 24, 26, 27, 33, 34
Montipora turgescens Bernard, 1897	13, 16, 18, 23, 25, 26, 27
Montipora turtlensis Veron and Wallace, 1984	17, 18
Montipora undata Bernard, 1897	23, 25, 28
Montipora venosa (Ehrenberg, 1834)	1, 4, 6, 8, 12, 14, 17, 21, 47
Montipora verrucosa (Lamarck, 1816)	2, 4, 6, 8, 10, 11, 12, 16, 17, 21, 22, 30, 31, 34, 38, 45, 54, 55
Montipora sp. 1	10, 13
Family Poritidae	
Alveopora catalai Wells, 1968	7, 11, 13, 27, 46, 56
Alveopora fenestrata Lamarck, 1816	3, 12
Alveopora spongiosa Dana, 1846	1, 5, 17, 18
Alveopora tizardi Bassett-Smith, 1890	15, 16, 18, 30, 38, 48, 49, 53

continued

Species	Site records
Goniopora columna Dana, 1846	9, 10, 11, 15, 21, 22, 23, 25, 29, 40, 51
Goniopora djiboutiensis Vaughan, 1907	8, 15
Goniopora fruticosa Saville-Kent, 1893	7, 8, 13, 18, 20, 42, 46
Goniopora lobata Edwards & Haime, 1860	1, 4, 7, 11, 12, 22, 26, 27
Goniopora minor Crossland, 1952	3, 4, 11, 14, 17, 22, 23, 27
Goniopora palmensis Veron and Pichon, 1982	15, 16, 22, 24
Goniopora pandoraensis Veron and Pichon, 1982	11, 25, 26
Goniopora somaliensis Vaughan, 1907	7
Goniopora stokesi Milne Edwards & Haime, 1851	7, 8
Goniopora stuchburyi Wells, 1955	1, 2, 8, 14, 15, 17, 22, 24, 26, 27, 28
Goniopora tenuidens (Quelch, 1886)	3, 19, 21, 23, 49, 55
Goniopora sp.1	4, 5, 9
Porites annae Crossland, 1952	3, 11, 15, 28
Porites attenuata Nemenzo, 1955	2, 3, 4, 7, 10, 11, 28, 29, 30, 31, 35, 48, 49, 53, 54, 57
Porites cumulatus Nemenzo, 1955	13, 22, 28, 36, 40, 42
Porites cylindrica Dana, 1846	1, 3, 4, 7, 8, 10, 11, 12, 13, 14, 15, 16, 17, 18, 20, 21, 22, 23, 24, 25, 26, 27, 28, 29, 30, 31, 32, 33, 37, 38, 39, 41, 42, 46, 47, 49, 50, 52, 53, 54, 55, 56, 57
Porites heronensis Veron, 1985	15
Porites horizontalata Hoffmeister, 1925	1, 3, 4, 5, 7, 10, 14, 17, 20, 22, 23, 25, 26, 29, 35, 38
Porites latistellata Quelch, 1886	23
Porites lichen Dana, 1846	1, 2, 4, 7, 8, 9, 10, 11, 12, 13, 14, 15, 16, 18, 20, 24, 25, 26, 28, 29, 40, 45, 48, 49, 50, 53, 54, 55
Porites lobata Dana, 1846	1, 2, 4, 5, 6, 7, 9, 10, 11, 13, 14, 17, 18, 20, 21, 22, 24, 25, 26, 27, 28
Porites lutea Milne Edwards & Haime, 1860	1, 2, 3, 4, 5, 6, 8, 9, 10, 11, 12, 13, 14, 15, 16, 17, 19, 20, 21, 22, 26, 27, 30, 33, 36, 37, 39, 42, 45, 46, 48, 49, 55
Porites mayeri Vaughan, 1918	10, 13, 20, 24
Porites monticulosa Dana, 1846	2, 22, 23, 24, 25, 38
Porites murrayensis Vaughan, 1918	8, 27
Porites nigrescens Dana, 1846	1, 3, 4, 5, 8, 12, 13, 14, 15, 17, 20, 21, 22, 38, 40, 45
Porites negrosensis Veron, 1990	9, 10
Porites rugosa Fenner & Veron, 2000	3, 7, 8, 9, 10, 11, 12, 13, 14, 16, 22, 25, 26, 29, 30, 34, 37, 40, 43, 48, 49, 50, 54, 57
Porites rus (Forsskål, 1775)	1, 2, 3, 4, 6, 9, 11, 12, 16, 17, 18, 19, 20, 21, 22, 23, 24, 25, 26, 27, 28, 30, 31, 35, 36, 37, 38, 48, 49, 50, 53, 54, 55, 56, 57
Porites solida (Forsskål 1775)	3, 8, 11, 12, 14, 16, 18, 20, 24, 26, 27
Porites vaughani Crossland, 1952	1, 3, 4, 5, 7, 8, 9, 10, 12, 13, 14, 16, 17, 18, 20, 23, 24, 25, 29, 37, 45
Porites flavus Veron, 2000	3, 4, 7, 12, 13, 14, 15, 20, 22
Porites sp.1	1, 4, 17, 19, 25
Porites sp.2	10, 12
Family Siderasteidae	
Coscinaraea columna (Dana, 1846)	1, 7, 8, 10, 11, 12, 14, 15, 16, 17, 18, 19, 24, 26, 29, 31, 32, 34, 36, 37, 39, 40, 41, 42, 43, 46, 47, 48, 49, 50, 51, 52, 53, 54, 55
Coscinaraea exesa (Dana, 1846)	10, 11

continued

Species	Site records
Coscinaraea monile (Forsskål, 1775)	8, 10, 12, 13, 41, 42, 57
Coscinaraea wellsi Veron and Pichon, 1980	1, 3, 16, 17, 18, , 27, 37, 52
Coscinaraea sp.1	8, 13, 14, 15, 20, 28, 34, 57
Psammocora contigua (Esper, 1797*)*	1, 3, 8, 11, 15, 22, 25, 26, 271, 3, 11, 15, 25, 26, 27, 29, 32, 35, 36, 38, 40, 41, 42, 46, 51, 57
Psammocora digitata Milne Edwards and Haime, 1851	1, 2, 3, 7, 8, 9, 10, 11, 12, 13, 14, 19, 20, 21, 22, 24, 25, 26, 28, 29, 30, 32, 33, 36, 40, 42, 46, 48, 54, 55, 56, 57
Psammocora explanulata van der Horst, 1922	1, 27, 49
Psammocora haimeana Milne Edwards & Haime, 1851	1, 3, 4, 10, 12, 14, 19
Psammocora neirstraszi van der Horst, 1921	1, 2, 3, 4, 5, 6, 14, 17, 19, 22, 23, 24, 25, 27, 28, 29, 37, 44, 48, 49, 52, 53, 57
Psammocora obtusangulata (Lamarck, 1816)	22
Psammocora profundacella Gardiner, 1898	1, 3, 6, 9, 10, 11, 12, 14, 16, 17, 18, 19, 23, 24, 25, 26, 27, 28, 31, 32, 35, 37, 39, 42, 44, 49, 53, 57
Psammocora superficialis Gardiner, 1898	9, 23, 25, 27, 28, 42
Psammocora vaughani Yabe & Sugiyama, 1936	41
Pseudosiderastrea tayami Yabe & Sugiyama, 1935	12
Family Agariciidae	
Coeloseris mayeri Vaughan, 1918	1, 2, 3, 4, 5, 7, 8, 10, 11, 12, 13, 14, 15, 16, 17, 18, 19, 20, 21, 22, 23, 26, 27, 28, 29, 30, 31, 32, 34, 35, 36, 37, 39, 41, 42, 43, 45, 46, 47, 48, 50, 52, 54, 55, 56, 57
Gardineroseris planulata (Dana, 1846)	1, 2, 3, 4, 5, 6, 9, 15, 18, 19, 20, 22, 23, 24, 25, 28, 29, 33, 35, 37, 39, 43, 44, 45, 46, 48, 49, 52, 53, 54, 55
Leptoseris explanata Yabe & Sugiyama, 1941	1, 2, 4, 5, 6, 7, 8, 9, 10, 12, 13, 15, 16, 18, 19, 20, 21, 22, 23, 24, 25, 26, 27, 28, 30, 33, 35, 37, 42, 44, 46, 49, 52, 56, 57
Leptoseris foliosa Dinesen, 1980	1, 2, 4, 6, 10, 11, 13, 15, 16, 19, 20, 22, 27, 28
Leptoseris gardineri Horst, 1921	13, 41, 56, 57
Leptoseris hawaiiensis Vaughan, 1907	2, 3, 4, 5, 6, 9, 10, 12, 13, 14, 15, 16, 17, 19, 20, 22, 23, 24, 25, 26, 29, 35, 37, 41, 52, 53, 55, 56, 57
Leptoseris incrustans (Quelch, 1886)	2, 53
Leptoseris mycetoseroides Wells, 1954	1, 2, 4, 6, 9, 10, 16, 17, 18, 19, 20, 22, 23, 24, 25, 26, 27, 28, 29, 32, 33, 35, 37, 38, 39, 43, 44, 46, 48, 49, 52, 53, 54, 56, 57
Leptoseris papyracea (Dana, 1846)	2, 25, 27
Leptoseris scabra Vaughan, 1907	2, 5, 6, 7, 9, 11, 12, 13, 14, 16, 18, 19, 20, 21, 22, 23, 24, 25, 26, 27, 29, 30, 32, 35, 37, 38, 49, 50, 52, 53
Leptoseris striata Fenner & Veron 2000	46, 53
Leptoseris cf. *tubulifera* Vaughan, 1907	13, 15, 16, 20, 22, 23, 24
Leptoseris yabei (Pillai & Scheer, 1976)	1, 5, 18, 20, 22, 26, 29, 32, 34, 37, 38, 39, 43, 50, 54
Leptoseris sp.1	16
Pachyseris foliosa Veron, 1990	8, 11, 23, 41, 42, 46
Pachyseris gemmae Nemenzo, 1955	4, 12, 13, 16, 17, 18, 22, 26, 28, 36, 56
Pachyseris sp.1	1
Pachyseris rugosa (Lamarck, 1801)	2, 3, 4, 11, 13, 14, 20, 22, 23, 25, 26, 27, 28, 30, 32, 33, 38, 39, 40, 42, 43, 46, 48, 50, 57
Pachyseris speciosa (Dana, 1846)	1, 2, 3, 4, 5, 6, 7, 8, 9, 10, 11, 12, 13, 14, 15, 17, 18, 19, 20, 22, 23, 24, 25, 26, 27, 28, 29, 30, 31, 33, 34, 35, 36, 38, 39, 40, 41, 43, 46, 48, 50, 51, 52, 53, 54, 55, 56, 57

continued

Species	Site records
Pavona bipartita Nemenzo, 1980	2, 16, 18, 30, 50, 52, 53, 57
Pavona cactus (Forsskål, 1775)	2, 3, 11, 22, 23, 24, 25, 26, 27, 28, 38, 42, 46, 48, 56, 57
Pavona clavus (Dana, 1846)	1, 2, 3, 4, 6, 8, 9, 11, 17, 24, 38, 39, 48, 54, 55, 56, 57
Pavona decussata (Dana, 1846)	1, 4, 5, 9, 11, 12, 15, 16, 18, 19, 21, 22, 23, 25, 26, 29, 30, 31, 32, 33, 34, 35, 37, 38, 39, 42, 43, 44, 46, 47, 48, 49, 53, 54, 55, 57
Pavona duerdeni Vaughan, 1907	3, 4, 6, 17, 28, 31, 39, 43, 44, 47, 48, 49, 52, 53, 54 , 55
Pavona explanulata (Lamarck, 1816)	2, 3, 4, 5, 6, 7, 9, 10, 12, 14, 15, 17, 18, 19, 20, 21, 22, 23, 24, 25, 26, 27, 28, 29, 31, 32, 34, 37, 38, 39, 41, 42, 43, 46, 48, 49, 52, 53, 55, 56, 57
Pavona frondifera (Lamarck, 1816)	15
Pavona minuta Wells, 1954	15, 18, 22, , 29, 37, 38, 44, 52, 54
Pavona varians Verrill, 1864	1, 2, 3, 4, 5, 6, 7, 8, 9, 10, 11, 12, 13, 14, 16, 17, 18, 19, 20, 21, 22, 23, 24, 25, 26, 27, 28, 29, 30, 31, 33, 35, 37, 38, 39, 42, 43, 44, 46, 47, 48, 49, 50, 52, 53, 54, 55, 57
Pavona venosa (Ehrenberg, 1834)	1, 3, 4, 5, 6, 8, 9, 11, 12, 14, 15, 16, 17, 18, 19, 20, 21, 22, 23, 24, 25, 26, 27, 31, 33, 38, 39, 43, 53, 54, 57
Pavona sp.1	11, 14, 21, 24
Pavona sp.2	1, 3, 4, 5, 8, 9, 10, 13, 16, 18, 19, 22, 23, 25, 27
Family Fungiidae	
Cantharellus jebbi Hoeksema, 1993	1, 3, 9, 12, 14, 15, 16, 19, 20, 23, 25, 37, 39, 49, 56
Cantharellus nuomeae Hoeksema & Best, 1984	2
Cantharellus sp.1	24
Ctenactis albitentaculata Hoeksema, 1989	1, 6, 7, 10, 11, 12, 13, 14, 15, 16, 17, 20, 24, 25, 26, 29, 30, 36, 38, 43, 45, 54, 55, 56, 57
Ctenactis crassa (Dana, 1846)	1, 2, 3, 6, 7, 8, 9, 10, 14, 16, 17, 18, 19, 20, 22, 23, 24, 25, 26, 27, 33, 36, 42, 46, 55
Ctenactis echinata (Pallas, 1766)	1, 3, 4, 5, 6, 7, 8, 9, 10, 11, 12, 13, 14, 15, 16, 17, 18, 20, 21, 22, 24, 25, 26, 27, 28, 30, 32, 33, 34, 35, 37, 38, 39, 41, 42, 43, 46, 48, 50, 54, 55, 57
Cycloseris colini Veron, 2000	15
Cycloseris costulata Ortmann, 1889	18, 27, 34
Cycloseris sinensis Milne Edwards and Haime, 1851	15, 57
Cycloseris somervillei (Gardiner, 1909)	2, 3, 9, 10, 13, 16, 17, 18, 25, 26, 28
Cycloseris vaughani (Boschma, 1923)	3, 27
Fungia concinna Verrill, 1864	1, 2, 3, 4, 13, 14, 15, 16, 23, 24, 25, 27, 28, 33, 34, 35, 48, 49, 50, 54, 55
Fungia danai (Milne Edwards & Haime, 1851)	11, 12, 25
Fungia fralinae Nemenzo, 1955	57
Fungia fungites (Linnaeus, 1758)	1, 2, 3, 4, 6, 7, 8, 9, 10, 11, 12, 14, 15, 16, 17, 18, 19, 20, 21, 22, 23, 24, 25, 26, 27, 29, 30, 32, 33, 34, 35, 36, 39, 43, 44, 48, 49, 55, 56
Fungia granulosa Klunzinger, 1879	2, 3, 5, 6, 8, 10, 11, 12, 13, 14, 15, 16, 17, 18, 19, 20, 21, 22, 24, 25, 26, 27, 28, 33, 34, 42, 50, 53, 55
Fungia horrida Dana, 1846	1, 2, 3, 4, 5, 6, 7, 8, 9, 10, 11, 12, 13, 14, 15, 16, 17, 18, 19, 20, 21, 22, 23, 24, 25, 26, 27, 28, 32, 33, 39, 49, 50, 54
Fungia klunzingeri Doderlein, 1901	1, 2, 3, 4, 5, 6, 7, 8, 9, 10, 11, 12, 13, 16, 18, 20, 23, 24, 25, 26, 27, 28, 29, 34, 36, 46, 57

continued

Species	Site records
Fungia moluccensis van der Horst, 1919	3, 4, 6, 11, 12, 15, 23, 24, 26, 27, 41, 57
Fungia paumotensis Stutchbury, 1833	1, 2, 3, 5, 6, 8, 9, 10, 12, 13, 14, 17, 18, 19, 20, 21, 22, 23, 24, 25, 26, 27, 30, 31, 32, 33, 34, 35, 36, 38, 39, 40, 41, 42, 43, 45, 46, 48, 49, 50, 51, 53, 54, 55, 57
Fungia repanda Dana, 1846	1, 2, 4, 5, 6, 7, 8, 9, 10, 11, 12, 13, 14, 15, 16, 17, 18, 19, 20, 21, 22, 23, 25, 26, 27, 28, 32, 33, 35, 36, 42, 46, 49, 53, 54, 55, 57
Fungia scabra Döderlein, 1901	10
Fungia scruposa Klunzinger, 1879	2, 4, 5, 6, 7, 8, 9, 16, 19, 22, 23, 26, 27, 32, 34, 36, 40, 42, 43, 49, 50, 57
Fungia scutaria Lamarck, 1801	4, 5, 6, 8, 9, 10, 17, 18, 19, 29, 33, 39, 43, 53, 55, 57
Fungia spinifer Claereboudt and Hoeksema, 1987	11, 12, 13, 48
Fungia sp.1	15
Halomitra clavator Hoeksema, 1989	17, 23, 57
Halomitra pileus (Linnaeus, 1758)	2, 3, 4, 5, 6, 7, 8, 9, 10, 11, 12, 13, 14, 15, 16, 17, 18, 19, 20, 21, 22, 23, 24, 25, 26, 27, 29, 30, 31, 32, 33, 34, 35, 36, 37, 39, 43, 46, 48, 49, 50, 54, 55, 56, 57
Heliofungia actiniformis Quoy and Gaimard, 1833	7, 10, 12, 15, 21, 22, 24, 25, 26, 27, 29, 31, 36, 43, 54, 56, 57
Herpolitha limax (Esper, 1797)	1, 2, 4, 5, 8, 11, 12, 13, 14, 15, 16, 17, 18, 20, 21, 22, 23, 24, 25, 26, 27, 28, 29, 32, 34, 35, 36, 41, 43, 46, 48, 50, 54, 56, 57
Herpolitha weberi Boschma, 1925	1, 2, 11, 14, 15, 22, 24, 57
Lithophyllon mokai Hoeksema, 1989	19, 23, 26
Podabacia crustacea (Pallas, 1766)	3, 6, 10, 11, 12, 21, 22, 23, 24, 25, 26, 51, 54
Podabacia motuporensis Veron, 1990	2, 6, 7, 8, 9, 11, 13, 15, 16, 17, 18, 19, 20, 21, 22, 23, 24, 25, 26, 27, 30, 34, 35, 36, 37, 39, 40, 41, 42, 43, 44, 46, 51, 52, 53, 54, 56, 57
Polyphyllia talpina (Lamarck, 1801)	8, 9, 11, 12, 13, 14, 15, 20, 21, 23, 26, 27, 28, 34, 40, 41, 46, 50, 55, 57
Sandalolitha dentata (Quelch, 1886)	1, 6, 9, 18, 19, 23, 29, 35, 37, 43, 48, 50, 53, 54
Sandalolitha robusta Quelch, 1886	3, 4, 7, 8, 9, 10, 11, 12, 14, 15, 18, 20, 22, 23, 24, 25, 26, 27, 31, 32, 33, 34, 36, 38, 39, 42, 43, 46, 48, 55, 56, 57
Zoopilus echinatus Dana, 1846	2, 7, 8, 9, 10, 12, 14, 16, 17, 19, 22, 25, 26, 27, 28, 30, 33, 35, 37, 43, 57
Family Oculinidae	
Galaxea horrescens (Dana, 1846)	3, 7, 10, 11, 13, 20, 21, 22, 25, 26, 31, 35, 38, 40, 48, 51
Galaxea astreata (Lamarck, 1816)	3, 5, 8, 11, 12, 15, 16, 17, 18, 21, 22, 23, 24, 25, 26, 34, 35, 36, 38, 42, 44, 45, 51, 52, 54
Galaxea fasicularis (Linnaeus, 1767)	1, 2, 3, 4, 5, 6, 7, 8, 9, 10, 11, 12, 13, 14, 15, 16, 17, 18, 19, 20, 21, 22, 23, 24, 25, 26, 27, 281, 29, 30, 31, 32, 33, 35, 36, 37, 38, 39, 41, 43, 46, 48, 49, 50, 51, 52, 53, 54, 55, 56, 57
Galaxea paucisepta Claereboudt, 1990	7, 15
Galaxea acrhelia Veron, 2000	13, 18, 20, 22, 23, 24, 25, 26, 28
Galaxea longisepta Fenner & Veron, 2000	1, 2, 4, 20, 22, 24, 25, 28, 57
Family Pectiniidae	
Echinophyllia aspera (Ellis & Solander, 1786)	1, 2, 3, 7, 8, 9, 10, 11, 12, 15, 16, 17, 19, 20, 21, 22, 23, 24, 25, 26, 28, 31, 32, 36, 39, 41, 42, 43, 46, 49, 53, 55, 56
Echinophyllia echinata (Saville-Kent, 1871)	6, 7, 8, 12, 15, 16, 19, 22, 26, 27, 29, 36, 37, 57
Echinophyllia echinoporoides Veron and Pichon, 1979	4, 9, 24, 26, 28
Echinophyllia orpheensis Veron & Pichon, 1979	10, 11, 12, 13, 20, 21, 24, 36, 41, 42, 51

continued

Species	Site records
Echinophyllia patula (Hodgson and Ross, 1982)	1, 12, 14, 16, 17, 18, 19, 20, 23, 24, 25, 26, 27, 28, 30, 33, 37, 38, 39, 43, 49, 52, 53, 55, 56, 57
Echinophyllia cf. *taylorae* (Veron, 2000)	7, 8, 11, 16, 28, 56
Mycedium elephantotus (Pallas, 1766)	1, 2, 4, 5, 6, 7, 8, 9, 10, 11, 12, 13, 14, 15, 16, 18, 19, 20, 21, 22, 24, 25, 26, 27, 28, 29, 30, 31, 32, 33, 34, 35, 36, 37, 38, 43, 44, 45, 48, 49
Mycedium robokaki Moll and Best, 1984	1, 2, 3, 4, 5, 6, 8, 9, 10, 11, 12, 15, 16, 17, 18, 19, 20, 22, 23, 25, 26, 27, 29, 30, 32, 35, 38, 40, 42, 43, 44, 49, 54
Mycedium mancaoi Nemenzo, 1979	20, 21, 22, 25
Oxypora crassispinosa Nemenzo, 1979	1, 14, 16, 17, 18, 19, 20, 22, 23, 24, 25, 26, 27, 28, 29, 35, 36, 37, 39, 55
Oxypora glabra Nemenzo, 1959	1, 5, 9, 11, 13, 18, 19, 22, 25, 27, 28, 35
Oxypora lacera (Verrill, 1864)	1, 2, 3, 4, 5, 6, 7, 8, 9, 10, 11, 12, 14, 15, 16, 17, 18, 19, 20, 21, 22, 23, 24, 25, 26, 27, 28, 29, 30, 31, 32, 33, 35, 36, 37, 39, 42, 43, 44, 46, 48, 49, 50, 52, 53, 55, 56, 57
Pectinia alcicornis (Saville-Kent, 1871)	1, 3, 7, 9, 10, 11, 12, 13, 14, 15, 16, 17, 21, 22, 23, 27, 34, 35, 38, 50, 51
Pectinia elongata Rehberg, 1892	21, 29, 51
Pectinia lactuca (Pallas, 1766)	3, 4, 5, 6, 9, 12, 16, 18, 19, 20, 21, 22, 23, 27, 28, 29, 30, 32, 34, 35, 37, 39, 42, 44, 46, 48, 49, 50, 52, 53, 54, 55, 56
Pectinia paeonia (Dana, 1846)	1, 3, 4, 7, 11, 12, 14, 16, 18, 20, 21, 22, 23, 24, 25, 26, 27, 28, 30, , 32, 33, 34, 35, 38, 40, 41, 42, 46, 49, 51, 54, 57
Pectinia cf. *maxima* (Moll and Best, 1984)	10, 12
Pectinia aylini (Wells, 1935)	1, 11, 14, 23, 27
Family Mussidae	
Micromussa amakusensis (Veron, 1990)	1, 7, 8, 26
Acanthastrea echinata (Dana, 1846)	1, 3, 7, 9, 10, 11, 12, 13, 14, 15, 16, 18, 20, 21, 23, 24, 26, 27, 29, 31, 32, 34, 35, 36, 37, 39, 44, 45, 47, 49, 50, 52, 53, 55, 57
Acanthastrea hemprichii (Ehrenberg, 1834)	1, 10
Acanthastrea hillae Wells, 1955	11
Acanthastrea rotundaflora Chevalier, 1975	8, 10, 11
Acanthastrea brevis Milne Edwards & Haime, 1849	15, 36, 43, 45, 48, 49
Acanthastrea faviaformis Veron, 2000	20, 36
Acanthastrea subechinata Veron, 2000	3, 11, 13, 20, 21
Acanthastrea sp. 1	3, 12, 15, 17, 18, 26
Acanthastrea sp.2	14, 18, 19, 20, 22, 23, 24, 26
Blastomussa merleti Wells, 1961	29, 46, 56, 57
Cynarina lacrymalis (Milne Edwards & Haime, 1848)	3, 10, 15, 23, 24, 46
Lobophyllia corymbosa (Forsskål, 1775)	2, 3, 8, 9, 11, 12, 15, 16, 18, 19, 21, 22, 24, 25, 26, 28, 40, 57
Lobophyllia hataii Yabe, Sugiyama & Eguchi 1936	1, 2, 3, 4, 5, 6, 11, 15, 17, 20, 21, 22, 23, 26, 27, 28, 34, 44, 46, 49, 55, 57
Lobophyllia hemprichii (Ehreberg, 1834)	2, 3, 4, 5, 6, 7, 8, 9, 10, 11, 12, 13, 14, 15, 17, 18, 19, 20, 21, 22, 23, 24, 25, 26, 27, 28, 29, 30, 31, 32, 33, 34, 35, 36, 37, 38, 39, 40, 41, 42, 43, 45, 46, 48, 49, 50, 51, 53, 55, 56
Lobophyllia pachysepta Chevalier, 1975	15, 34, 42, 51
Lobophyllia robusta Yabe, Sugiyama & Eguchi 1936	1, 2, 3, 4, 5, 6, 8, 9, 12, 13, 14, 15, 17, 19, 20, 21, 22, 23, 26, 27, 28, 29, 30, 33, 34, 36, 40, 43, 44, 45, 46, 48, 49, 52, 53, 55, 56, 57

continued

Species	Site records
Lobophyllia dentata Veron, 2000	25, 32, 33, 34, 55, 56
Lobophyllia flabelliformis Veron, 2000	4, 13, 53
Lobophyllia serratus Veron, 2000	21
Scolymia australis (Milne Edwards and Haime, 1849)	1, 2, 5, 15, 18, 19, 26
Scolymia vitiensis Brüggemann, 1877	4, 5, 6, 8, 9, 10, 11, 12, 13, 15, 16, 18, 19, 20, 21, 22, 23, 24, 25, 26, 27, 29, 32, 33, 34, 36, 37, 41, 42, 44, 46, 49, 52, 53, 55, 57
Symphyllia agaricia Milne Edwards & Haime, 1849	1, 2, 3, 4, 6, 19, 24, 25, 28, 29, 45, 49, 53, 55
Symphyllia hassi Pillai and Scheer, 1976	1, 3, 14, 16, 17, 18, 21, 24, 25, 26, 34, 42
Symphyllia radians Milne Edwards and Haime, 1849	1, 2, 4, 6, 7, 8, 9, 10, 12, 13, 14, 15, 17, 19, 20, 21, 22, 25, 26, 27, 28, 29, 30, 31, 32, 33, 34, 35, 36, 39, 41, 43, 44, 45, 46, 47, 48, 49, 52, 53, 54, 55, 56, 57
Symphyllia recta (Dana, 1846)	1, 2, 3, 4, 6, 7, 8, 9, 10, 11, 12, 14, 15, 16, 17, 18, 19, 20, 21, 22, 23, 24, 25, 26, 27, 28, 31, 33, 35, 37, 39, 43, 44, 47, 48, 50, 51, 54
Symphyllia valenciennesi Milne Edwards & Haime, 1849	1, 6, 14, 18, 19, 20, 23, 26, 29, 31, 42
Family Merulinidae	
Hydnophora exesa (Pallas, 1766)	2, 4, 7, 8, 9, 11, 12, 13, 14, 15, 16, 18, 19, 20, 22, 25, 26, 28, 29, 30, 32, 35, 36, 37, 38, 39, 43, 44, 46, 47, 49, 50, 51, 53, 54, 55, 57
Hydnophora grandis Gardiner, 1904	7, 8, 13, 14, 16, 17, 19, 20, 21, 27, 29, 32, 33, 34, 40, 42, 46, 48, 51, 53, 54, 55, 57
Hydnophora microconos (Lamarck, 1816)	4, 5, 6, 7, 8, 11, 12, 14, 15, 16, 17, 18, 19, 20, 22, 23, 25, 27, 28, 29, 30, 31, 33, 36, 37, 39, 43, 44, 47, 48, 49, 52, 53, 55, 57
Hydnophora pilosa Veron, 1985	2, 5, 7, 8, 32, 39, 56
Hydnophora rigida (Dana, 1846)	1, 2, 3, 4, 5, 6, 7, 8, 9, 10, 11, 12, 13, 14, 15, 16, 17, 19, 21, 22, 23, 24, 25, 26, 27, 28, 29, 30, 32, 33, 35, 37, 38, 39, 43, 48, 50, 51, 53, 54, 55, 57
Merulina ampliata (Ellis and Solander, 1786)	1, 3, 4, 6, 7, 8, 9, 10, 11, 12, 13, 14, 15, 16, 17, 18, 19, 20, 21, 22, 23, 24, 25, 26, 27, 28, 29, 31, 33, 34, 35, 36, 37, 38, 39, 41, 42, 43, 44, 46, 47, 48, 49, 50, 51, 53, 55, 56, 57
Merulina scabricula Dana, 1846	1, 3, 4, 7, 8, 9, 10, 11, 12, 13, 14, 15, 16, 17, 18, 19, 20, 21, 22, 23, 24, 25, 26, 27, 29, 30, 31, 32, 34, 35, 39, 43, 46, 48, 49, 53, 54, 56
Paraclavarina triangularis Veron and Pichon, 1979	7, 10, 15, 21, 32, 34, 42, 46, 51, 56, 57
Scapophyllia cylindrica Milne Edwards and Haime, 1848	1, 2, 3, 5, 7, 14, 15, 19, 27, 42, 53
Family Faviidae	
Australogyra zelli (Veron, Pichon, and Wijsman-Best, 1977)	7, 12, 32, 33, 34, 46
Caulastrea echinulata (Milne Edwards and Haime, 1849)	2, 11, 12, 14, 20, 25, 37, 38, 39, 56
Caulastrea furcata Dana, 1846	11, 12, 14, 28, 51
Cyphastrea agassizi (Vaughan, 1907)	17, 25, 30
Cyphastrea chalcidicum (Forsskål, 1775)	8, 12, 13, 16, 17, 19, 21, 22, 23, 26, 27
Cyphastrea decadia Moll and Best, 1984	10, 11, 12, 13, 14, 15, 16, 17, 20, 22, 24, 25, 29, 30, 33, 38, 40, 48, 49, 50
Cyphastrea japonica Yabe and Sugiyama, 1932	24
Cyphastrea microphthalma (Lamarck, 1816)	3, 4, 6, 7, 8, 10, 12, 16, 17, 19, 20, 22
Cyphastrea serailia (Forsskål, 1775)	1, 2, 3, 4, 5, 6, 7, 8, 10, 11, 12, 13, 14, 16, 17, 18, 19, 20, 21, 22, 24, 26, 27
Cyphastrea sp. 1	12

continued

Species	Site records
Cyphastrea sp. 2	19, 24
Diploastrea heliopora (Lamarck, 1816)	1, 2, 3, 4, 5, 6, 7, 8, 9, 10, 11, 12, 13, 14, 15, 16, 17, 18, 19, 20, 21, 22, 23, 24, 25, 26, 27, 28, 29, 30, 31, 32, 33, 34, 35, 36, 37, 38, 39, 40, 41, 42, 43, 44, 45, 46, 47, 48, 49, 50, 51, 53, 54, 55, 56, 57
Echinopora gemmacea (Lamarck, 1816)	1, 4, 6, 8, 9, 10, 12, 13, 14, 15, 16, 17, 18, 19, 20, 21, 22, 23, 24, 25, 26, 27, 28, 29, 30, 32, 33, 34, 35, 36, 38, 39, 42, 43, 44, 45, 46, 48, 49, 50, 52, 53, 54, 55, 56, 57
Echinopora hirsutissima Milne Edwards and Haime, 1849	1, 3, 4, 5, 6, 7, 8, 14, 16, 17, 18, 19, 22, 23, 25, 27, 28, 29, 30, 31, 34, 35, 36, 39, 43, 44, 45, 48, 49, 50, 52, 53, 55
Echinopora horrida Dana, 1846	1, 3, 6, 7, 8, 10, 11, 12, 13, 14, 15, 16, 17, 19, 20, 21, 22, 23, 24, 25, 26, 27, 28, 29, 30, 32, 33, 35, 36, 39, 46, 48, 54, 55, 56, 57
Echinopora lamellosa (Esper, 1775)	1, 2, 3, 4, 6, 7, 8, 9, 10, 11, 12, 13, 14, 15, 16, 17, 19, 20, 21, 22, 23, 24, 25, 26, 27, 29, 35, 37, 55
Echinopora mammiformis (Nemenzo, 1959)	3, 4, 8, 10, 11, 12, 13, 14, 15, 16, 20, 21, 22, 23, 28, 29, 30, 31, 32, 33, 34, 35, 36, 37, 38, 39, 40, 42, 43, 46, 48, 49, 51, 54, 55, 56, 57
Echinopora pacificus Vernon, 1990	7, 11, 16, 56
Favia danae Verrill, 1872	11, 12
Favia favus (Forsskål, 1775)	1, 2, 3, 4, 6, 7, 8, 9, 10, 11, 12, 13, 15, 17, 19, 21, 22, 25, 27
Favia laxa (Klunzinger, 1879)	21, 24, 36, 51
Favia lizardensis Veron, Pichon & Wijsman-Best, 1972	3, 7, 8, 10, 11, 12, 14, 15, 16, 20, 21, 22, 23, 24, 25, 26, 27, 28, 29, 32, 34, 36, 50, 54, 57
Favia maritima (Nemenzo, 1971)	7, 11, 12, 20, 26
Favia matthaii Vaughan, 1918	1, 2, 3, 4, 5, 6, 8, 10, 11, 12, 15, 16, 17, 18, 19, 20, 21, 22, 23, 24, 25, 26, 27, 28, 29, 32, 46, 49, 51, 57
Favia maxima Veron, Pichon & Wijsman-Best, 1972	15
Favia pallida (Dana, 1846)	3, 4, 5, 6, 7, 8, 9, 10, 11, 12, 14, 15, 16, 17, 18, 19, 20, 21, 22, 23, 24, 25, 26, 27, 28, 30
Favia rotundata Veron, Pichon & Wijsman-Best, 1972	2, 7, 9, 10, 12, 15, 20, 21, 26, 31, 33, 34, 51, 52
Favia speciosa (Dana, 1846)	3, 7, 12, 14, 15, 20, 21, 23, 24, 26, 27, 33
Favia stelligera (Dana, 1846)	1, 2, 3, 4, 5, 6, 7, 8, 9, 10, 11, 12, 13, 14, 15, 16, 17, 18, 19, 20, 21, 22, 23, 24, 25, 26, 27, 28, 29, 30, 31, 32, 33, 34, 35, 36, 38, 39, 41, 42, 43, 44, 45, 46, 47, 48, 49, 50, 52, 53, 54, 55
Favia truncatus Veron, 2000	1, 4, 6, 8, 11, 12, 15, 17, 20, 27, 33, 40, 47
Favites abdita (Ellis & Solander, 1786)	1, 3, 4, 5, 7, 8, 10, 11, 12, 15, 18, 19, 20, 21, 22, 23, 24, 26, 27, 29, 30, 31, 32, 33, 34, 35, 36, 44, 47, 48, 49, 50, 51, 52, 54, 55, 57
Favites chinensis (Verrill, 1866)	5, 13
Favites complanata (Ehrenberg, 1834)	1, 2, 3, 4, 5, 6, 7, 8, 9, 10, 11, 12, 13, 15, 17, 18, 20, 21, 22, 25, 26, 27, 28
Favites flexuosa (Dana, 1846)	4, 5, 7, 8, 10, 12, 13, 19, 23
Favites halicora (Ehrenberg, 1834)	1, 3, 5, 7, 8, 11, 14, 16, 17, 18, 21, 24, 25, 29, 33, 34, 36, 50, 53
Favites paraflexuosa Veron, 2000	5, 11, 12, 14, 15
Favites pentagona (Esper, 1794)	1, 2, 5, 7, 12, 15, 25, 28
Favites rosaria Veron, 2000	2, 8, 14, 18
Favites russelli (Wells, 1954)	1, 4, 9, 12, 13, 15, 16, 17, 18, 28, 33
Favites vasta Klunzinger, 1879	7, 10, 15, 16, 18, 22, 28

continued

Species	Site records
Goniastrea aspera Verrill, 1905	1, 12, 15, 19, 22, 25, 26, 32, 34, 35, 36, 38, 41, 43, 48, 49, 57
Goniastrea australensis (Milne Edwards & Haime, 1857)	1, 15, 19, 21, 23, 27
Goniastrea edwardsi Chevalier, 1971	1, 2, 3, 4, 5, 6, 7, 8, 9, 10, 11, 12, 14, 15, 16, 17, 18, 19, 20, 21, 22, 23, 25, 26, 27, 28, 29, 31, 32, 33, 34, 35, 36, 37, 39, 40, 41, 42, 44, 46, 47, 48, 49, 50, 51, 52, 54, 55, 57
Goniastrea favulus (Dana, 1846)	15
Goniastrea pectinata (Ehrenberg, 1834)	1, 2, 3, 4, 5, 7, 8, 9, 10, 11, 12, 13, 14, 16, 17, 18, 19, 20, 21, 22, 23, 24, 25, 26, 27, 28, 29, 30, 31, 33, 35, 36, 37, 38, 39, 43, 44, 45, 48, 49, 50, 51, 53, 54, 55, 57
Goniastrea retiformis (Lamarck, 1816)	1, 3, 4, 5, 6, 7, 8, 10, 12, 14, 15, 16, 18, 19, 21, 22, 25, 26, 27, 28, 29, 30, 31, 32, 33, 34, 36, 37, 38, 39, 43, 44, 45, 46, 48, 49, 50, 52, 53, 54, 55
Leptastrea bewickensis Veron, Pichon, and Wijsman-Best, 1977	8, 26
Leptastrea inaequalis Klunzinger, 1879	7, 19, 29, 30, 31, 32, 39, 43, 44, 47, 48, 49, 50, 53, 55
Leptastrea pruinosa Crossland, 1952	1, 3, 4, 5, 6, 9, 10, 11, 12, 14, 15, 16, 17, 18, 19, 22, 23, 26, 27, 31, 34
Leptastrea purpurea (Dana, 1846*)*	1, 2, 3, 4, 6, 7, 10, 11, 12, 13, 15, 16, 17, 18, 19, 20, 21, 22, 23, 24, 25, 26, 27, 28, 29, 30, 32, 33, 35, 37, 38, 39, 40, 41, 42, 43, 45, 46, 47, 49, 50, 51, 54, 57
Leptastrea transversa Klunzinger, 1879	1, 2, 3, 4, 7, 8, 9, 10, 11, 12, 13, 15, 16, 17, 18, 19, 20, 21, 24, 25, 26, 28, 29, 31, 37, 39, 43, 44, 47, 48, 49, 52, 53, 55
Leptoria irregularis Veron, 1990	4, 5, 10, 18, 24
Leptoria phrygia (Ellis & Solander, 1786)	1, 2, 3, 4, 5, 6, 7, 9, 11, 12, 13, 14, 15, 16, 17, 18, 19, 21, 22, 24, 25, 26, 27, 28, 29, 31, 32, 33, 34, 35, 36, 37, 38, 39, 41, 42, 43, 44, 45, 47, 48, 49, 50, 52, 53, 54, 55, 56, 57
Montastrea annuligera (Milne Edwards and Haime, 1849)	1, 2, 5, 7, 8, 9, 10, 11, 12, 15, 17, 18, 19, 20, 21, 26, 27, 28
Montastrea curta (Dana, 1846)	1, 4, 5, 7, 8, 12, 13, 15, 16, 17, 18, 19, 20, 21, 24, 25, 26, 29, 30, 31, 32, 33, 34, 37, 39, 43, 44, 45, 46, 47, 48, 49, 50, 52, 53, 54, 55
Montastrea magnistellata Chevalier, 1971	1, 2, 4, 5, 7, 10, 11, 12, 15, 21, 23, 24, 26, 27, 31, 33, 34, 36, 37, 49, 51, 54
Montastrea salebrosa (Nemenzo, 1959)	15, 21
Montastrea valenciennesi (Milne Edwards and Haime, 1848)	1, 4, 10
Montastrea multipunctata Hodgson, 1985	4, 8, 11, 21
Oulastrea crispata Lamarck, 1816	30, 31, 52
Oulophyllia bennettae (Veron, Pichon and Wijsman-Best, 1977)	1, 3, 6, 10, 13, 17, 21, 23, 28, 34, 46, 48, 51
Oulophyllia crispa (Lamarck, 1816)	1, 4, 5, 6, 7, 8, 10, 11, 12, 13, 14, 15, 17, 18, 19, 20, 21, 22, 23, 24, 25, 26, 27, 27, 29, 31, 32, 33, 34, 36, 39, 40, 41, 42, 43, 44, 45, 46, 48, 49, 51, 52, 53, 54, 55, 56
Oulophyllia levis Nemenzo, 1959	26, 28
Platygyra contorta Veron, 1990	9, 10, 14, 15, 17, 18, 19, 20, 22, 23, 25, 27
Platygyra daedalea (Ellis & Solander, 1786)	1, 2, 3, 4, 5, 6, 7, 8, 9, 10, 11, 12, 13, 14, 15, 16, 17, 18, 19, 20, 21, 22, 23, 24, 25, 26, 27, 28, 29, 30, 31, 32, 33, 34, 35, 36, 43, 44, 45, 46, 47, 49, 50, 52, 53, 54, 55, 57
Platygyra lamellina (Ehrenberg, 1834)	2, 3, 4, 5, 6, 7, 9, 11, 13, 14, 15, 17, 19, 20, 22, 23, 24, 25, 26, 27, 28, 29, 32, 33, 34, 36, 37, 39, 43, 49, 50, 51, 57
Platygyra pini Chevalier, 1975	3, 4, 9, 11, 12, 13, 16, 17, 18, 19, 21, 22, 23, 25, 26, 27, 53, 56
Platygyra ryukyuensis Yabe and Sugiyama, 1936	11, 12, 14, 15, 26
Platygyra sinensis (Milne Edwards & Haime, 1849)	1, 2, 3, 4, 5, 6, 7, 8, 10, 12, 14, 15, 17, 18, 23, 24, 26, 27, 48

continued

Species	Site records
Platygyra verweyi Wijsman-Best, 1976	7, 8, 12, 14, 15, 17
Platygyra yaeyemaensis Eguchi and Shirai, 1977	2, 5, 10, 14, 17
Platygyra acuta Veron, 2000	1, 7, 8, 12, 14, 15, 17, 21, 28
Plesiastrea versipora (Lamarck, 1816)	2, 5, 12, 18, 22, 24, 37, 38, 39, 42, 43, 44, 53, 55, 56, 57
Family Trachyphyllidae	
Trachyphyllia goeffroyi (Audouin, 1826)	15
Family Euphyllidae	
Catalaphyllia jardinei (Saville-Kent, 1893)	23
Euphyllia ancora Veron and Pichon, 1979	4, 25, 43, 56
Euphyllia cristata Chevalier, 1971	1, 4, 6, 7, 9, 10, 11, 15, 18, 21, 23, 24, 25, 26, 27, 30, 32, 35, 42, 45, 49, 51, 57
Euphyllia glabrescens (Chamisso and Eysenhardt, 1821)	3, 4, 8, 9, 15, 16, 17, 19, 20, 23, 24, 25, 26, 27, 49
Euphyllia yaeyamensis (Shirai, 1980)	8, 11, 22, 23, 26, 54
Physogyra lichtensteini Milne Edwards and Haime, 1851	2, 3, 4, 5, 6, 7, 9, 10, 11, 12, 13, 14, 15, 16, 17, 18, 19, 20, 21, 22, 23, 24, 25, 26, 27, 28, 29, 30, 31, 34, 35, 36, 37, 38, 39, 40, 41, 42, 43, 44, 45, 46, 48, 49, 51, 52, 53, 54, 55, 56, 57
Plerogyra simplex Rehberg, 1892	4, 13, 15, 23, 35, 36, 46, 51
Plerogyra sinuosa (Dana, 1846)	1, 2, 4, 9, 12, 15, 22, 24, 26, 29, 35, 36, 42, 53, 57
Family Dendrophylliidae	
Balanophyllia sp. 1	52, 55
Dendrophyllia sp. 1	1, 31, 49, 52, 54, 55
Rhizopsammia verrilli van der Horst, 1922	4, 22, 24, 27, 31, 38, 44, 45, 50, 52, 57
Turbinaria frondens (Dana, 1846)	22, 37, 47, 54
Turbinaria mesenterina (Lamarck, 1816)	2, 3, 4, 5, 8, 9, 11, 12, 13, 16, 17, 18, 19, 21, 22, 24, 26, 29, 30, 33, 35, 36, 37, 38, 40, 43, 47, 48, 49, 50, 51, 55, 56
Turbinaria peltata (Esper, 1794)	4, 9, 15, 18, 19, 25, 27, 28, 31, 36, 37, 39, 40, 41, 42
Turbinaria reniformis Bernard, 1896	1, 2, 3, 4, 5, 6, 7, 8, 10, 11, 12, 13, 14, 15, 17, 18, 19, 20, 22, 23, 27, 28, 29, 30, 31, 32, 35, 37, 38, 39, 40, 42, 43, 44, 45, 46, 47, 48, 49, 50, 52, 53, 54, 55, 56, 57
Turbinaria stellulata (Lamarck, 1816)	4, 5, 7, 11, 12, 13, 14, 17, 18, 19, 20, 22, 27, 31, 35, 37, 38, 39, 43, 45, 46, 47, 49, 55, 56
Tubastraea coccinea Lesson, 1829	4, 37, 52
Tubastraea micrantha Ehrenberg, 1834	1, 4, 5, 9, 18, 19, 22, 24, 31, 33, 35, 37, 39, 47, 48, 52, 54, 55
Subclas Alcyonaria (Octocorallia)	
Order Alcyonacea	
Family Tubiporidae	
Tubipora musica Linnaeus, 1758	15, 26, 29, 30, 35, 43, 45, 46, 47, 48, 49, 52, 53
Tubipora sp. 1	1, 13, 18, 25, 31, 37, 43, 50, 54, 55, 57
Tubipora sp. 2	4, 34, 50
Order Helioporacea (Coenothecalia)	
Family Helioporidae	
Heliopora coerulea (Pallas, 1766)	11, 12, 42, 47, 48, 53, 54, 56
Class Hydrozoa	
Order Milleporina	
Family Milleporidae	
Millepora dichotoma Forsskål, 1775	8, 11, 12, 17, 18, 19, 28, 39, 43

continued

Species	Site records
Millepora cf. *exaesa* Forsskål, 1775	1, 2, 3, 4, 6, 11, 20, 21, 25, 27, 29, 31, 33, 35, 37, 38, 39, 45, 47, 52, 57
Millepora intricata Milne-Edwards and Haime, 1857	1, 3, 8, 11, 14, 20, 23, 27, 42, 53, 56, 57
Millepora tenera Boschma, 1949	47
Millepora sp. 1	5, 6, 8, 9, 14, 16, 17, 18, 28, 49
Order Sylasterina	
Family Stylasteridae	
Distichopora nitida Verrill, 1864	31, 36, 50, 54, 55
Distichopora violacea (Pallas, 1766)	30, 44, 52, 53
Stylaster sp. 1	3, 4, 6, 9, 12, 31, 37, 38, 44, 48, 50, 52, 53, 54, 55, 57

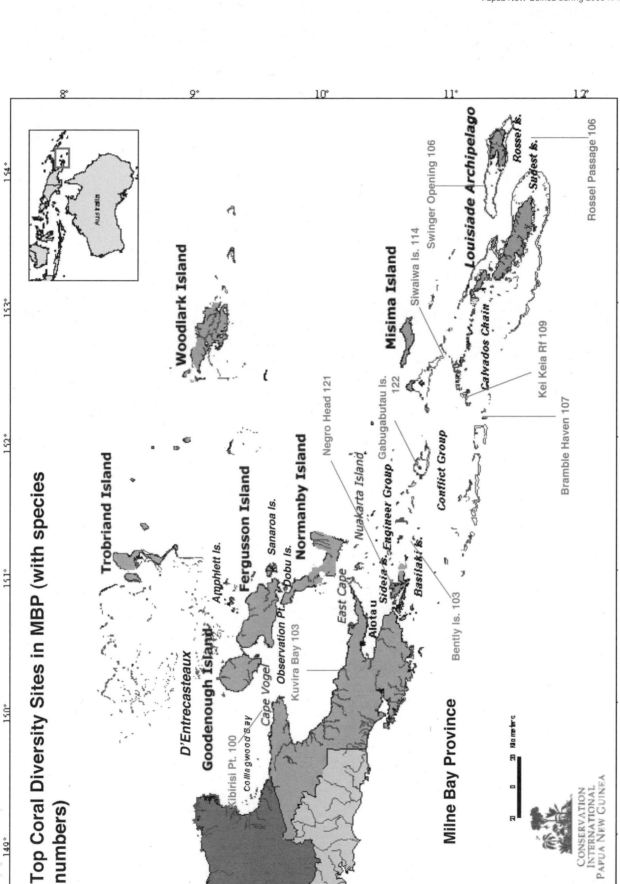

Top Coral Diversity Sites in MBP (with species numbers)

Trobriand Island

Woodlark Island

Amphlett Is.

D'Entrecasteaux

Goodenough Island

Fergusson Island

Sanaroa Is.

Dobu Is.

Normanby Island

Kibirisi Pt. 100

Collingwood Bay

Cape Vogel

Observation Pt.

Kuvira Bay 103

East Cape

Alotau

Sideia Is.

Nuakarta Island

Engineer Group

Gabugabutau Is. 122

Basilaki Is.

Negro Head 121

Bently Is. 103

Conflict Group

Bramble Haven 107

Kei Keia Rf 109

Siwaiwa Is. 114

Swinger Opening 106

Misima Island

Calvados Chain

Louisiade Archipelago

Sudest Is.

Rossel Is.

Rossel Passage 106

Milne Bay Province

Australia

CONSERVATION
INTERNATIONAL
PAPUA NEW GUINEA

Kilometers

Appendix 2

Data for calculating the Reef Condition Index

Gerald R. Allen, Pamela Seeto, and Tessa McGarry

Site	Coral spp.	Fish spp.	Cond. pts.	RC.	Site	Coral spp.	Fish spp.	Cond. pts.	RC.
1	70	160	90	151.90	30	78	232	140	206.49
2	58	171	200	196.60	31	78	201	160	204.51
3	78	179	140	187.50	32	95	190	150	209.12
4	77	260	180	233.91	33	90	169	110	184.20
5	61	194	170	193.48	34	93	154	140	190.17
6	59	224	160	198.11	35	98	224	130	214.48
7	50	193	100	158.17	36	91	173	90	172.69
8	63	189	120	170.51	37	93	203	190	231.12
9	64	194	160	191.24	38	88	191	130	195.00
10	51	211	110	164.59	39	100	224	200	252.43
11	91	188	160	209.87	40	54	162	140	163.03
12	91	210	190	233.51	41	44	141	130	143.29
13	53	182	120	161.83	42	81	198	90	173.93
14	72	243	80	179.23	43	100	223	180	243.71
15	80	194	110	180.82	44	70	230	180	217.81
16	84	153	130	178.33	45	57	238	170	206.12
17	80	177	170	202.03	46	92	172	130	191.29
18	100	180	140	206.99	47	53	167	140	164.03
19	87	208	170	218.48	48	114	210	180	244.51
20	80	197	150	204.69	49	108	211	170	235.71
21	71	183	180	202.50	50	84	200	170	215.40
22	97	218	150	220.63	51	62	192	150	184.43
23	64	140	110	149.23	52	69	187	180	201.68
24	67	168	100	156.99	53	91	198	180	222.53
25	97	162	140	196.11	54	107	223	180	244.48
26	90	210	70	178.19	55	103	163	180	219.98
27	104	177	80	179.58	56	87	203	190	225.79
28	74	170	130	176.73	57	120	209	190	253.31
29	121	233	180	258.82					

Appendix 3

Percentage of various substrata for Milne Bay survey sites

Gerald R. Allen, Pamela Seeto, and Tessa McGarry

The letters "s" and "d" refer to the shallow (approximately 8 m) and deep (approximately 16 m) transects at the same site.

Site	Hard corals	Dead coral	Rubble	Soft corals	Sand	Sponge	Algae
1s	19	36	17	0	3	16	9
1d	16	23	21	6	1	29	4
2	77	1	16	0	1	5	0
3	40	7	7	4	19	10	13
4s	57	1	5	7	3	8	19
4d	52	0	3	9	10	15	11
5	56	1	8	23	3	9	0
6s	43	0	38	18	0	0	1
6d	30	1	41	24	2	2	0
7s	32	3	6	2	37	14	6
7d	28	1	2	2	38	28	1
8s	85	0	6	5	1	3	0
8d	59	0	5	5	16	13	2
9s	34	0	37	21	3	2	3
9d	43	0	5	40	2	8	2
10s	49	2	17	4	23	5	0
10d	46	2	15	4	17	16	0
11s	52	2	26	0	0	19	1
11d	60	0	9	1	11	19	0
12s	61	2	5	7	12	13	0
12d	56	0	4	5	8	26	1
13s	65	8	9	4	7	7	0
13d	68	4	4	4	2	18	0
14	39	4	16	17	2	4	18
15s	47	11	2	8	23	7	2
15d	43	2	0	17	3	18	17
16s	52	5	23	17	1	2	0

continued

Site	Hard corals	Dead coral	Rubble	Soft corals	Sand	Sponge	Algae
16d	44	5	26	20	1	3	1
17s	66	2	9	2	0	16	5
17d	35	3	20	10	6	32	4
18s	48	1	2	36	0	11	2
18d	33	0	11	21	5	30	0
19s	70	3	22	0	1	4	0
19d	57	0	20	1	10	12	0
20s	42	2	15	4	16	21	0
20d	32	2	13	5	7	38	3
21	30	3	19	2	30	16	0
22s	37	0	19	1	13	29	1
22d	38	2	5	3	24	28	0
23s	24	5	16	1	29	25	0
23d	3	1	0	0	78	18	0
24s	54	7	13	0	2	23	1
24d	35	8	6	4	6	37	4
25s	57	5	20	7	0	11	0
25d	51	3	5	5	0	36	0
26s	36	14	0	3	36	11	0
26d	24	6	12	2	35	21	0
27s	77	5	10	2	1	5	0
27d	30	7	37	3	14	9	0
28s	72	9	8	8	2	1	0
28d	71	3	20	0	0	6	0
29s	54	0	7	24	13	0	2
29d	54	0	2	23	2	7	12
30s	34	7	30	13	2	3	11
30d	39	1	33	11	0	5	11
31s	22	0	24	10	39	3	2
31d	22	0	12	23	35	8	0
32s	38	4	29	1	24	1	3
32d	47	2	14	0	28	8	1
33s	33	5	36	1	4	14	7
33d	46	3	30	0	19	2	0
34s	53	9	3	0	9	26	0
34d	32	0	2	1	43	22	0
35s	48	0	30	6	5	8	3
35d	30	1	0	18	39	10	2
36s	20	4	55	11	10	0	0
36d	27	12	33	10	5	13	0
37s	67	2	2	17	0	4	8
37d	53	0	7	20	7	6	7
38s	29	1	8	8	39	9	6

continued

Site	Hard corals	Dead coral	Rubble	Soft corals	Sand	Sponge	Algae
38d	25	0	0	18	40	17	0
39s	56	1	3	6	31	0	3
39d	43	1	0	8	47	1	0
40	33	1	0	4	13	3	46
41s	22	1	21	1	3	5	47
41d	13	1	15	11	4	0	56
42s	41	2	5	9	33	3	7
42d	25	2	3	16	40	5	9
43s	76	0	2	2	2	2	16
43d	36	1	19	12	12	13	7
44	7	0	0	3	0	2	88
45s	24	8	9	1	17	13	28
45d	15	0	52	3	6	3	21
46s	52	2	15	21	1	6	3
46d	37	2	2	27	29	3	0
47s	52	0	4	15	21	8	0
47d	14	0	0	0	60	26	0
48s	25	4	8	48	11	3	1
48d	42	1	5	23	19	10	0
49s	28	3	18	7	27	9	8
49d	19	0	3	2	66	10	0
50s	13	3	30	12	36	2	4
50d	33	1	38	13	5	8	2
51s	35	1	1	2	54	7	0
51d	26	0	1	6	54	9	4
52s	55	0	0	12	0	2	31
52d	27	0	0	50	0	21	2
53s	48	2	0	32	3	15	0
53d	34	0	0	55	0	11	0
54s	52	1	5	4	36	2	0
54d	45	0	16	23	13	3	0
55s	55	0	24	7	13	1	0
55d	32	0	52	11	0	4	1
56s	70	0	8	16	4	2	0
56d	30	0	4	10	30	18	8
57s	67	0	4	12	0	17	0
57d	48	0	13	8	9	22	0

Appendix 4

List of molluscs recorded at Milne Bay, Papua New Guinea during RAP surveys

Fred E. Wells

Site records are for the second (2000) survey. Species lacking site numbers were collected only on the 1997 survey. Summary of top molluscs diversity sites are presented in the map following the table.

Species	Site records
CLASS POLYPLACOPHORA	
Family Cryptoplacidae	
Cryptoplax larvaeformis (Burrow, 1815)	11, 13
Family Chitonidae	
Acanthopleura gemmata (Blainville, 1825)	4, 15, 22, 24, 26, 27
CLASS GASTROPODA	
Family Patellidae	
Cellana sp.	
Cellana rota (Gmelin, 1791)	27
Cellana tramoserica (Holten, 1802)	
Patella flexuosa Quoy & Gaimard, 1834	11
Patelloida saccharina (Linnaeus, 1758)	13
Patelloida striata (Quoy & Gaimard, 1834)	7
Family Haliotidae	
Haliotis asinina Linnaeus, 1758	12
Haliotis crebrisculpta Sowerby, 1914	21
Haliotis glabra Gmelin, 1791	10, 27
Haliotis ovina Gmelin, 1791	4, 7, 13, 19, 21, 27, 28
Haliotis planata Sowerby, 1833	6, 10, 15
Haliotis cf. *pulcherrima* Gmelin, 1791	19
Haliotis varia Linaeus, 1758	27
Family Fissurellidae	
Diodora mus (Reeve, 1850)	
Diodora galeata (Helbling, 1779)	9
Diodora ticaonica (Reeve, 1850)	
Suctus granulatus Blainville, 1819	
Family Turbinidae	
Astralium rhodostomum (Lamarck, 1822)	7, 8, 17

Species	Site records
Astralium sp.	7, 10, 12
Monodonta labio (Linnaeus, 1758)	
Turbo argyrostomus (Linnaeus, 1758)	2, 4, 10, 15-20, 23, 26,27
Turbo bruneus (Röding, 1798)	1
Turbo chrysostoma (Linnaeus, 1758)	11
Turbo cinereus Born, 1778	
Turbo crassus Wood, 1828	
Turbo foliaceus Philippi, 1847	
Turbo petholatus Linnaeus, 1758	1-4, 7, 8, 11-13, 15, 18, 20, 22, 24-28
Turbo setosus Gmelin, 1791	
Family Trochidae	
Angaria delphinus (Linnaeus, 1758)	
Calthalotia attenuata (Jonas, 1844)	12
Calthalotia sp.	7
Chrysostoma paradoum (Born, 1780)	10, 13
Clanculus atropurpureus (Gould, 1849)	10, 13
Clanculus clanguloides (Wood, 1828)	8
Emarginula incisura Adams, 1851	
Ethminolia stearnsi (Pilsbry, 1895)	6
Euchelus atratus (Gmelin, 1791)	12
Gena impertusa (Burrow, 1815)	
Liotina peronii (Kiener, 1839)	
Monilea belcheri (Philippi, 1849)	
Pseudostomatella decolorata (Gould, 1848)	10
Tectus fenestratus Gmelin, 1790	3, 15, 18, 21
Tectus maculatus Linnaeus, 1758	4, 7, 21
Tectus pyramis Born, 1778	4-9, 12, 13, 15-18, 22, 24-28

continued

Species	Site records
Tectus triserialis (Lamarck, 1822)	12
Thalotia attenuata (Jonas, 1844)	
Trochus hanleyanus (Reeve, 1843)	7
Trochus histrio Reeve, 1842	1
Trochus niloticus Linnaeus, 1767	4, 7, 11-13, 17, 21, 22, 28
Trochus stellatus Gmelin, 1791	6
Stomatella auricula (Lamarck, 1816)	10
Stomatia phymotis Helbling, 1779	12
Umbonium vestiarum (Linnaeus, 1758)	1,12
Family Neritopsidae	
Neritopsis radula Gray, 1842	12, 13, 16, 19, 24
Family Neritidae	
Nerita albicilla Linnaeus, 1758	7, 10, 11
Nerita costata Gmelin, 1791	7, 27
Nerita grossa Linnaeus, 1758	27
Nerita plicata Linnaeus, 1758	10, 11, 22, 27
Nerita polita Linnaeus, 1758	4, 7, 25-27
Nerita reticulata Karsten, 1789	
Nerita undata Linnaeus, 1758	4, 7, 11, 13, 24-27
Neritina violacea (Gmelin, 1791)	27
Theodoxus oualiensis (Lesson, 1831)	
Family Cerithiidae	
Cerithium alveolus Hombron & Jacquinot, 1854	7, 13
Cerithium balteatum Philippi, 1848	26
Cerithium citrinum Sowerby, 1855	4, 5, 8
Cerithium columna Sowerby, 1834	5, 7-9, 18, 24, 25, 27
Cerithium echinatum (Lamarck, 1822)	2, 4, 5, 6
Cerithium lifuense Melvill & Standen, 1895	1, 9, 22, 23, 25-27, 28
Cerithium munitum Sowerby, 1855	21, 22, 27
Cerithium nesioticum Pilsbry & Vanetta, 1906	5, 6, 7, 9, 11, 17-19, 22
Cerithium nodulosus (Bruguière, 1792)	4, 15
Cerithium punctatum Bruguière, 1792	10, 11
Cerithium rostratum Sowerby, 1855	10, 12, 14
Cerithium salebrosum Sowerby, 1855	3, 7, 12, 18, 19, 21
Cerithium tenuifilosum Sowerby, 1866	3, 6, 9, 15, 17, 20
Cerithium torresi Smith, 1884	3
Cerithium trailli (Sowerby, 1855)	13
Cerithium zonatum (Wood, 1828)	
Clypeomorus batillariaeformis Habe & Kosuge, 1966	
Clypeomorus moniliferum (Kiener, 1841)	15
Clypeomorus petrosum (Wood, 1828)	6

Species	Site records
Rhinoclavis articulata (Adams & Reeve, 1850)	13
Rhinoclavis aspera (Linnaeus, 1758)	4-6, 8, 10-15, 17-23, 26-28
Rhinovlavis fasciatus (Bruguière, 1792)	1, 3, 5, 10, 19, 21
Rhinoclavis longicaudata (Adams & Reeve, 1850)	15
Rhinoclavis sinensis (Gmelin, 1791)	1, 6, 8, 10, 11, 12, 15, 18, 19, 27
Rhinoclavis vertagus (Linnaeus, 1767)	3
Family Turritellidae	
Gazemeda declivis (Adams & Reeve, 1848)	
Family Planaxidae	
Planaxis decollatus (Quoy & Gaimard, 1834)	
Planaxis niger Quoy & Gaimard, 1834	11
Planaxis sulcatus (Born, 1780)	10
Family Potamididae	
Cerithidea cingulata (Gmelin, 1791)	
Terebralia palustris (Linnaeus, 1767)	13, 14
Terebralia sulcata (Born, 1778)	13
Telescopium telescopium (Linnaeus, 1758)	
Family Modulidae	
Modulus tectum (Gmelin, 1791)	17, 21, 23, 27
Family Littorinidae	
Littorina coccinea (Gmelin, 1791)	7, 11, 27
Littoraria scabra (Linnaeus, 1758)	
Littorina undulata Gray, 1839	7
Nodilittorina millegrana (Philippi, 1848)	7, 11
Nodilittorina pyramidalis (Quoy & Gaimard, 1833)	
Tectarius pagodus (Linnaeus, 1758)	27
Family Rissoidae	
Rissoid sp. 1	
Rissoid sp. 2	
Zebina gigantea (Deshayes, 1850)	19
Family Strombidae	
Lambis chiragra (Linnaeus, 1758)	10, 13
Lambis lambis (Linnaeus, 1758)	4
Lambis millepeda (Linnaeus, 1758)	1, 3, 7, 8, 15-18, 21, 25, 27
Lambis scorpius (Linnaeus, 1758)	1, 6, 15-17, 24, 27
Lambis truncata (Humphrey, 1786)	13, 15, 18
Strombus aurisdianae Linnaeus, 1758	13

continued

Species	Site records
Strombus dentatus Linnaeus, 1758	12, 20, 28
Strombus dilatatus (Swainson, 1821)	
Strombus epidromis Linnaeus, 1758	3
Strombus erythrinus Dillwyn, 1817	22
Strombus fragilis (Röding, 1798)	
Strombus gibberulus Linnaeus, 1758	3, 4, 7, 10-14, 21, 23, 27
Strombus labiatus (Röding, 1798)	18
Strombus lentiginosus Linnaeus, 1758	12
Strombus luhuanus Linnaeus, 1758	3-5, 7, 10-14, 18, 22, 26, 28
Strombus microurceus (Kira, 1959)	4, 11-14
Strombus minimus Linnaeus, 1771	3, 4, 7, 8, 14, 19, 27
Strombus mutabilis Swainson, 1821	1, 11
Strombus plicatus (Röding, 1798)	14
Strombus sinuatus Humphrey, 1786	
Strombus terebellatus Sowerby, 1842	
Strombus urseus Linnaeus, 1758	3, 10, 15, 18, 19, 21, 23
Strombus variabilis Swainson, 1820	14
Strombus vomer (Röding, 1798)	
Terebellum terebellum (Linnaeus, 1758)	3, 7, 8, 10-12, 20, 27, 28
Family Vanikoridae	
Vanikoro cancellata (Lamarck, 1822)	17, 18, 27
Family Hipponicidae	
Hipponix conicus (Schumacher, 1817)	2, 4, 6, 9, 13, 14, 16, 18, 19, 20, 28
Family Calyptraeidae	
Calyptraea calyptraeformis (Lamarck, 1822)	
Crepidula aculeata (Gmelin, 1791)	
Family Capulidae	
Capulus liberatus Pease, 1868	3, 7, 9, 11, 18, 22
Cheilea equestris (Linnaeus, 1758)	2, 3, 8, 21, 27, 28
Cheilea sp. 2	27
Family Vermetidae	
Serpulorbis colubrina (Röding, 1798)	1, 2, 5, 7, 11, 12, 16, 18, 20, 25-27 ,28
Family Cypraeidae	
Cypraea annulus Linnaeus, 1758	4, 7, 10, 13, 15, 18, 27
Cypraea arabica Linnaeus, 1758	13, 15, 22, 27
Cypraea argus Linnaeus, 1758	6, 12, 21
Cypraea asellus Linnaeus, 1758	10, 15, 18, 22, 28
Cypraea becki Gaskoin, 1836	12

Species	Site records
Cypraea brevidentata (Sowerby, 1820)	
Cypraea caputserpentis Linnaeus, 1758	4, 8, 10, 13, 27
Cypraea carneola Linnaeus, 1758	1, 6, 17, 19, 20, 23
Cypraea caurica Linnaeus, 1758	
Cypraea cicercula Linnaeus, 1758	4, 6, 8, 9, 13
Cypraea clandestina Linnaeus, 1767	
Cypraea cribraria Linnaeus, 1758	
Cypraea cylindrica Born, 1778	25
Cypraea eglantina (Duclos, 1833)	4, 5, 8, 10, 18
Cypraea erosa Linnaeus, 1758	6, 11, 12, 21-23, 26
Cypraea fimbriata Gmelin, 1791	6, 8, 16
Cypraea flaveola Linnaeus, 1758	6, 9, 23, 27, 28
Cypraea gracilis Gaskoin, 1849	27
Cypraea globulus Linnaeus, 1758	6
Cypraea helvola Linnaeus, 1758	
Cypraea hirundo Linnaeus, 1758	6
Cypraea isabella Linnaeus, 1758	4, 6, 8, 9, 12, 15, 17, 18, 19, 21, 24-28
Cypraea kieneri Hidalgo, 1906	
Cypraea labrolineata Gaskoin, 1848	18, 20, 21, 27
Cypraea limacina Lamarck, 1810	
Cypraea lynx Linnaeus, 1758	1-5, 7, 8, 11, 12, 15, 17, 20-23, 26-28
Cypraea mappa Linnaeus, 1758	27
Cypraea mauritania Linnaeus, 1758	
Cypraea microdon Gray, 1828	10
Cypraea moneta Linnaeus, 1758	4, 7, 10-13, 15, 18, 20, 21, 23, 27
Cypraea nucleus Linnaeus, 1758	9, 20, 22, 28
Cypraea pallidula Gaskoin, 1849	18
Cypraea punctata Linnaeus, 1758	1, 9, 18, 27
Cypraea quadrimaculata Gray, 1824	11, 17, 20, 21
Cypraea staphylaea Linnaeus, 1758	4, 19
Cypraea talpa Linnaeus, 1758	4, 6, 10
Cypraea teres Gmelin, 1791	14, 22
Cypraea testudinaria Linnaeus, 1758	22, 27
Cypraea tigris Linnaeus, 1758	3, 4, 6, 17, 26
Cypraea ursellus Gmelin, 1791	28
Cypraea vitellus Linnaeus, 1758	5
Family Ovulidae	
Calpurneus lacteus (Lamarck, 1810)	
Calpurneus verrucosus (Linnaeus, 1758)	
Crenavolva striatula (A. Adams, 1855)	15
Ovula ovum (Linnaeus, 1758)	5
Phenacovolva sp.	13

continued

Species	Site records
Testudovula nebula (Azuma & Cate, 1971)	
Family Triviidae	
Trivia oryza (Lamarck, 1810)	1, 7, 10, 11, 15, 18, 21, 27
Family Lamellariidae	
Chelynotus tonganus (Quoy & Gaimard, 1832)	
Lamellariid sp. 1	
Family Naticidae	
Eunaticina linneana (Rècluz, 1843)	27
Natica gualtieriana (Rècluz, 1844)	
Natica lineozona Jousseaume, 1874	20
Natica onca (Röding, 1798)	3
Natica violacea Sowerby, 1825	5, 7
Natica vitellus (Linnaeus, 1758)	
Polinices aurantius (Röding, 1798)	
Polinices maurus (Lamarck, 1816)	
Polinices melanostomus (Gmelin, 1791)	5, 10, 13, 23
Polinices simae (Deshayes, 1838)	
Polinices tumidus (Swainson, 1840)	3, 5, 8, 10, 11, 12, 15, 23
Family Bursidae	
Bursa cruentata (Sowerby, 1835)	7
Bursa granularis (Röding, 1798)	4,10
Bursa lamarckii (Deshayes, 1853)	1, 6, 8, 9, 12, 17, 18, 20, 24, 25, 28
Bursa leo Shikama, 1964	8, 13
Bursa mammata (Röding, 1798)	
Bursa nobilis (Reeve, 1844)	3
Bursa rhodostoms Sowerby, 1840	
Bursa rosa Perry, 1811	6
Bursa tuberossissima (Reeve, 1844)	
Tutufa bubo (Linnaeus, 1758)	
Tutufa rubeta (Linnaeus, 1758)	2
Tutufa tenuigranosa Smith, 1914	
Family Cassidae	
Casmaria erinaceus (Linnaeus, 1758)	13, 26
Cassis cornuta (Linnaeus, 1758)	3
Phalium bisulcatum (Schubert & Wagner, 1829)	
Family Ranellidae	
Charonia tritonis (Linnaeus, 1758)	5
Cymatium aquatile (Reeve, 1844)	9
Cymatium flaveolum (Röding, 1798)	24

Species	Site records
Cymatium gutturnium (Röding, 1798)	15
Cymatium moritinctum (Reeve, 1844)	
Cymatium mundum (Gould, 1849)	13
Cymatium nicobaricum (Röding, 1798)	5
Cymatium pileare (Linnaeus, 1758)	
Cymatium pyrum (Linnaeus, 1758)	21
Cymatium rubeculum (Linnaeus, 1758)	7-9, 16, 17, 21, 24, 25, 28
Cymatium succinctum (Linnaeus, 1771)	27
Distorsio anus (Linnaeus, 1758)	7
Gyrineum bituberculare (Lamarck, 1816)	
Gyrineum gyrineum (Linnaeus, 1758)	7, 12, 20
Gyrineum pusillum (A. Adams, 1854)	3
Gyrineum roseum (Reeve, 1844)	
Linatella succincta (Linnaeus, 1771)	27
Ranularia muricinum (Gmelin, 1791)	
Septa gemmata (Reeve, 1844)	18, 21
Family Tonnidae	
Malea pomum (Linnaeus, 1758)	22
Tonna chinensis (Dillwyn, 1817)	
Tonna galea (Linnaeus, 1758)	
Tonna perdix (Linnaeus, 1758)	5, 21
Family Triphoridae	
Triphora sp. 1	19
Triphora sp. 2	10
Triphora sp. 3	7
Family Epitoniidae	
Epitonium perplexum (Pease, 1860)	
Family Muricidae	
Aspella anceps (Lamarck, 1822)	
Chicoreus banksii (Sowerby, 1841)	9
Chicoreus brunneus (Link, 1810)	3, 9, 18, 28
Chicoreus cumingii (A. Adams, 1853)	18
Chicoreus microphyllus (Lamarck, 1816)	4
Coralliophila costularis (Lamarck, 1816)	11
Coralliophila erosa (Röding, 1798)	
Coralliophila galea (Reeve, 1846)	26
Coralliophila neritoidea (Lamarck, 1816)	1, 2, 4-12, 15-28
Coralliophila pyriformis Kira, 1954	
Cronia fenestrata (Blainville, 1832)	4
Cronia funiculus (Wood, 1828)	10, 15
Cronia margariticola (Broderip, 1833)	
Drupa grossularia (Röding, 1798)	4, 6, 7, 9, 10, 11, 12, 18, 22, 25

continued

Species	Site records
Drupa morum (Röding, 1798)	4, 8 27
Drupa ricinus (Linnaeus, 1758)	6, 8, 12, 13, 15, 24
Drupa rubusidaeus (Röding, 1798)	18, 19, 25, 27
Drupella cariosa (Wood, 1828)	10, 11, 15, 16, 23, 26, 27
Drupella cornus (Röding, 1798)	1, 2, 5-10, 12, 16-20, 22-28
Drupella ochrostoma (Blainville, 1832)	1-3, 5, 7-9, 12, 16-18, 24-28
Drupella rugosa (Born, 1778)	20
Favartia sp.	
Homalocantha anomalae Kosuge, 1979	
Homalocantha anatomica (Perry, 1811)	6, 27
Homalocantha zamboi (Burch & Burch, 1960)	27
Latiaxis sp. 1	
Latiaxis sp. 2	
Maculotriton sculptile (Reeve, 1844)	
Maculotriton serriale (Deshayes, 1831)	17
Morula anaxeres (Kiener, 1835)	1, 6, 7, 8, 15, 17, 22
Morula aurantiaca (Hombron & Jacquinot, 1853)	
Morula biconica (Blainville, 1832)	10
Morula dumosa (Conrad, 1837)	28
Morula granulata (Duclos, 1832)	7, 10, 13, 22, 24, 26, 27
Morula margariticola (Broderip, 1832)	
Morula musiva (Kiener, 1836)	13
Morula nodicostata (Pease, 1868)	
Morula nodulifera (Menke, 1829)	10
Morula parva (Reeve, 1846)	
Morula spinosa (H. & A. Adams, 1855)	1, 3, 6, 8, 9, 11, 15, 17, 22
Morula uva (Röding, 1798)	2, 6, 9, 15, 16, 18-21
Murex ramosus (Linnaeus, 1758)	
Murex tenuirostrum Lamarck, 1822	14
Muricodrupa fiscella (Gmelin, 1791)	
Muricodrupa stellaris (Hombron & Jacquinot, 1853).	18
Muricodrupa sp.	
Muricopsis sp.	8
Naquetia triquetra (Born, 1778)	
Nassa serta (Bruguière, 1789)	9
Pterynotus barclayanus (A. Adams, 1873)	
Pterynotus martinetana (Röding, 1798)	6
Quoyola madreporarum (Sowerby, 1832)	5, 11, 13, 18, 27
Rapa rapa (Gmelin, 1791)	18, 22

Species	Site records
Thais aculeata (Desyayes, 1844)	
Thais armigera (Link, 1807)	
Thais echinata Blainville, 1832	
Thais kieneri (Deshayes, 1844)	
Thais mancinella (Linnaeus, 1758)	
Thais savignyi (Deshayes, 1844)	25, 27
Thais tuberosa (Röding, 1798)	4, 8, 13
Family Turbinellidae	
Vasum ceramicum (Linnaeus, 1758)	4, 7, 10-13, 16, 18, 28
Vasum turbinellus (Linnaeus, 1758)	7, 16, 18, 22, 25
Family Buccinidae	
Colubraria antiquata (Hinds, 1844)	25
Colubraria castanea Kuroda & Habe, 1952	6, 21
Colubraria muricata (Lightfoot, 1786)	
Colubraria nitidula (Sowerby, 1833)	18, 27
Colubraria tortuosa (Reeve, 1844)	
Colubraria sp.	12
Cantharus fragaria (Wood, 1828)	5, 6, 7, 27, 28
Cantharus fumosus (Dillwyn, 1817)	5, 26, 27, 28
Cantharus iostomus (Gray in Griffith & Pidgeon, 1834)	5, 10, 19, 20, 22
Cantharus pulcher (Reeve, 1846)	1, 2, 5, 7, 9, 12, 18, 26
Cantharus subrubiginosus (E.A. Smith, 1879)	2
Cantharus undosus (Linnaeus, 1758)	7, 11, 15, 21
Cantharus wagneri (Anton, 1839)	1, 6, 7, 21, 26
Cantharus wrightae Cernohorsky, 1974	27
Engina alveolata (Kiener, 1836)	10
Engina concinna (Reeve, 1846)	1
Engina incarnata (Deshayes, 1834)	6, 12, 19
Engina lineata (Reeve, 1846)	4, 8, 10, 12, 13, 18, 21, 22
Engina mendicaria (Linnaeus, 1758)	4
Engina zonalis (Lamarck, 1822)	2, 10
Maculotriton sculptile (Deshayes, 1834)	
Phos roseatus (Hinds, 1844)	14
Phos sculptilis Watson, 1886	1, 3
Phos textum (Gmelin, 1791)	1, 3,1 0-12, 22
Pisania fasciculata (Reeve, 1846)	
Pisania gracilis (Reeve, 1846)	3
Pisania ignea (Gmelin, 1791)	10
Pisania truncata (Hinds, 1844)	6, 10

continued

Species	Site records
Family Columbellidae	
Aeosopus spiculum (Duclos, 1846)	
Anachis miser (Sowerby, 1844)	13
Mitrella albina (Kiener, 1841)	1, 3, 10, 13, 15, 27
Mitrella ligula (Duclos, 1840)	3, 4, 10, 11, 15, 18, 19, 21, 26, 28
Mitrella livescens (Reeve, 1859)	27
Mitrella marquesa (Gaskoin, 1852)	
Mitrella puella (Sowerby, 1844)	14
Mitrella sp.	6
Pyrene deshayesii (Crosse, 1859)	11
Pyrene flava (Bruguière, 1789)	4, 15, 27
Pyrene ocellata (Link, 1807)	
Pyrene punctata (Bruguière, 1789)	9, 19
Pyrene scripta (Lamarck, 1822)	7
Pyrene splendidula (Sowerby, 1844)	1
Pyrene testudinaria (Link, 1807)	11, 13, 18, 27
Pyrene turturina (Lamarck, 1822)	1, 2, 4, 6, 7, 9-13, 15-28
Pyrene varians (Sowerby, 1832)	
Family Nassariidae	
Hebra horrida (Dunker, 1847)	3, 4, 7, 11, 23, 27
Nassarius albescens (Dunker, 1846)	3, 4, 7, 10-12, 22, 27
Nassarius cinctellus (Gould, 1850)	3, 21 ,26
Nassarius concinnus (Powys, 1835)	
Nassarius crematus (Hinds, 1844)	14
Nassarius delicatus (A. Adams, 1852)	14
Nassarius distortus (Adams, 1852)	
Nassarius dorsatus (Röding, 1798)	
Nassarius gaudiosus (Hinds, 1844)	13
Nassarius graniferus (Kiener, 1834)	5, 7, 12, 20, 21, 28
Nassarius livescens (Philippi, 1849)	14
Nassarius luridus (Gould, 1850)	
Nassarius multipunctatus (Schepman, 1911)	25
Nassarius pauperus (Gould, 1850)	1, 6
Nassarius reevanus (Dunker, 1847)	13
Nassarius splendidulus (Dunker, 1846)	14
Family Fasciolariidae	
Dolicholatirus lancea (Gmelin, 1791)	10, 17
Latirolagena smaragdula (Linnaeus, 1758)	
Latirus amplustris (Dillwyn, 1817)	7, 12, 21, 27
Latirus belcheri (Reeve, 1847)	17, 20
Latirus blosvillei (Deshayes, 1832)	

Species	Site records
Latirus craticularis (Linnaeus, 1758)	
Latirus gibbulus (Gmelin, 1791)	2, 9, 16, 17, 19, 27
Latirus nodatus (Gmelin, 1791)	1, 2, 6, 9, 16-20, 25, 27
Latirus noumeensis (Crosse, 1870)	21, 27
Latirus paetelianus (Kobelt, 1876)	3, 18, 20, 23, 26, 27
Latirus pictus (Reeve, 1847)	18, 27
Latirus polygonus (Gmelin, 1791)	26
Latirus turritus (Gmelin, 1791)	1, 4, 6, 8, 16-18, 20-23, 25, 27
Peristernia cf. *philberti* (Récluz, 1844)	7
Peristernia hesterae Melvill, 1911	22
Peristernia incarnata (Deshayes, 1830)	18
Peristernia lirata (Pease, 1868)	21
Peristernia nassatula (Lamarck, 1822)	13, 19, 27
Peristernia ustulata (Reeve, 1847)	21
Pleuroploca filamentosa (Röding, 1798)	2, 4, 6, 10, 16, 18, 19, 25
Pleuroploca trapezium (Linnaeus, 1758)	
Family Volutidae	
Cymbiola aulica (Sowerby, 1825)	
Cymbiola rutila (Broderip, 1826)	
Cymbiola vespertilio (Linnaeus, 1758)	
Melo sp.	
Family Olividae	
Oliva annulata (Gmelin, 1791)	5-8, 10, 12, 15, 18-22, 28
Oliva caerulea (Röding, 1798)	
Oliva caldania (Duclos, 1835)	
Oliva carneola (Gmelin, 1791)	3, 4, 7, 10, 11, 14, 23, 27
Oliva erythrostoma Meuschen, 1787	
Oliva hirasei Kira, 1959	4
Oliva miniacea Röding, 1798	2, 14
Olive mustellina Lamarck, 1811	14
Oliva oliva Linnaeus, 1758	3, 4, 7, 14
Oliva parkinsoni Prior, 1975	
Oliva paxillus Reeve, 1850	4, 5, 19
Oliva reticulata (Röding, 1798)	
Oliva tessellata Lamarck, 1811	14
Olivella sp. 1	
Olivella sp. 2	
Olivella sp. 3	
Family Harpidae	
Harpa amouretta Röding, 1798	10

continued

Species	Site records
Harpa articularis Lamarck, 1822	
Harpa major Röding, 1798	3,4
Family Mitridae	
Cancilla filaris (Linnaeus, 1771)	10, 11
Cancilla granatina (Lamarck, 1811)	
Cancilla praestantissima (Röding, 1798)	
Imbricaria conovula (Quoy & Gaimard, 1833)	10
Imbricaria conularis (Lamarck, 1811)	
Imbricaria olivaeformis (Swainson, 1821)	10, 12, 18, 19, 23
Imbricaria punctata (Swainson, 1821)	8, 15, 22, 24
Mitra acuminata Swainson, 1824	2, 9
Mitra amaura Hervier, 1897	6
Mitra assimilis Pease, 1868	9
Mitra aurantia (Gmelin, 1791)	
Mitra cardinalis (Gmelin, 1791)	
Mitra chrysostoma Broderip, 1836	
Mitra coarctata Reeve, 1844	2
Mitra contracta Swainson, 1820	3, 12, 16, 17, 20, 22, 24
Mitra coronata Lamarck, 1811	12
Mitra cucumerina Lamarck, 1811	
Mitra decurtata Reeve, 1844	4
Mitra eremitarum Röding, 1798	15, 22, 23
Mitra ferruginea Lamarck, 1811	25
Mitra fraga (Quoy & Gaimard, 1833)	12, 19, 22, 23
Mitra litterata Lamarck, 1811	
Mitra luctuosa A. Adams, 1853	7
Mitra mitra (Linnaeus, 1758)	1
Mitra nubila (Gmelin, 1791)	11, 21
Mitra paupercula Linnaeus, 1758	27
Mitra peculiaris (Reeve, 1845)	5, 11
Mitra retusa Lamarck, 1811	
Mitra rosacea Reeve, 1845	11
Mitra rubritincta Reeve, 1844	
Mitra scutulata (Gmelin, 1791)	
Mitra stictica (Link, 1807)	
Mitra tabanula Lamarck, 1811	25
Mitra telescopium Reeve, 1844	12 ,20, 25
Mitra turgida Reeve, 1845	
Mitra typha Reeve, 1845	22
Neocancilla clathrus (Gmelin, 1791)	11, 12, 21
Neocancilla papilio (Link, 1807)	12, 13, 22
Pterygia fenestrata (Lamarck, 1811)	

Species	Site records
Pterygia scabricula (Linnaeus, 1758)	
Pusia patriarchilis (Gmelin, 1791)	
Scabricola desetangsii (Kiener, 1838)	1
Family Costellariidae	
Vexillum amabile (Reeve, 1845)	10
Vexillum amanda (Reeve, 1845)	
Vexillum aureolineatum Turner, 1988	
Vexillum bernhardina (Röding, 1798)	11
Vexillum cadaverosum (Reeve, 1844)	3, 8, 23
Vexillum consangiuneum (Reeve, 1844)	11
Vexillum coronatum (Helbling, 1779)	2, 3, 10
Vexillum costatum (Gmelin, 1791)	18
Vexillum crocatum (Lamarck, 1811)	
Vexillum daedelum (Reeve, 1845)	
Vexillum deshayesii (Reeve, 1844)	
Vexillum echinatum (A. Adams, 1853)	3, 7, 21
Vexillum exasperatum (Gmelin, 1791)	3, 7, 14, 21, 22
Vexillum granosum (Gmelin, 1791)	22
Vexillum interruptum (Anton, 1838)	19
Vexillum lucidum (Reeve, 1845)	3
Vexillum lyratum (Lamarck, 1811)	1, 14
Vexillum macrospirum (A. Adams, 1853)	7
Vexillum michaui (Crosse & Fischer, 1854)	14
Vexillum mirabile (A. Adams, 1853)	10
Vexillum modestum (Reeve, 1845)	
Vexillum pacificum (Reeve, 1845)	5, 7, 10, 18, 19
Vexillum patriarchalis (Gmelin, 1791)	26
Vexillum plicarium (Linnaeus, 1758)	22
Vexillum polygonum (Gmelin, 1791)	2, 3
Vexillum radix (Sowerby, 1874)	24
Vexillum regina (Swainson, 1828)	
Vexillum semicostatum (Anton, 1838)	11
Vexillum subdivisum (Gmelin, 1791)	
Vexillum taeniatum (Lamarck, 1811)	
Vexillum turben (Reeve, 1844)	
Vexillum tankervillei (Melvill, 1888)	
Vexillum turrigerum (Reeve, 1845)	3
Vexillum unifascialis (Lamarck, 1811)	1, 5, 11, 21
Vexillum unifasciatum (Wood, 1828)	1
Vexillum vulpecula (Linnaeus, 1758)	14
Vexillum sanguisugum (Linnaeus, 1758)	21, 22, 26, 27
Vexillum zebuense (Reeve, 1845)	
Vexillum zelotypum Reeve, 1845	7

continued

Species	Site records
Family Cancellariidae	
Cancellaria elegans Sowerby, 1821	
Family Turridae	
Clavus bilineatus (Reeve, 1845)	19
Clavus canalicularis (Röding, 1798)	3, 6, 7
Clavus exasperatus (Reeve, 1843)	6
Clavus flammulatus (Montfort, 1810)	
Clavus laetus (Hinds, 1843)	19
Clavus lamberti (Montrouzier, 1860)	1, 6, 9
Clavus pica (Reeve, 1843)	
Clavus sp. 1	3, 21, 27
Clavus unizonalis (Lamarck, 1822)	11
Comitas cf. *kamakurana* (Pilsbry, 1895)	26
Daphnella cf. *aureola* (Reeve, 1845)	6
Epirirona sp.	
Eucithara conohelicoides (Reeve, 1846)	13
Eucithara reticulata (Reeve, 1846)	
Gemmula sp.	3, 10
Gemmula congener (E.A. Smith, 1894)	3, 11
Gemmula graffei (Weinkauff, 1875)	22
Gemmula kieneri (Doumet, 1840)	11
Gemmula monilifera (Pease, 1860)	3, 14
Inquisitor sp.	
Inquisitor sterrha (Watson, 1881)	2, 3
Inquisitor varicosa (Reeve, 1843)	3
Lienardia sp.	14, 27
Lophiotoma abbreviata (Reeve, 1843)	
Lophiotoma acuta (Perry, 1811)	
Lophiotoma indica (Röding, 1798)	3-5
Philbertia ??	9, 16
Turridrupa bijubata (Reeve, 1843)	10
Turridrupa cerithina (Anton, 1839)	18, 22, 27
Turris babylonia (Linnaeus, 1758)	
Turris crispa (Lamarck, 1816)	3, 23
Turris spectabilis (Reeve, 1843)	
Vexitomina regia (Reeve, 1842)	
Xenoturris cingulifera (Lamarck, 1822)	3, 22
Family Terebridae	
Duplicaria evoluta (Deshayes, 1859)	
Hastula albula (Menke, 1843)	22
Hastula lanceata (Linnaeus, 1767)	1
Hastula solida (Deshayes, 1857)	5
Hastula strigilata (Linnaeus, 1758)	
Terebra affinis Gray, 1834	1, 8
Terebra areolata (Link, 1807)	11, 14, 21

Species	Site records
Terebra babylonia Lamarck, 1822	22
Terebra cerithina Lamarck, 1822	1
Terebra chlorata Lamarck, 1822	
Terebra cingulifera Lamarck, 1822	
Terebra columellaris Hinds, 1844	2, 8, 19
Terebra conspersa Hinds, 1844	14
Terebra crenulata (Linnaeus, 1758)	23
Terebra cumingi Deshayes, 1857	4
Terebra dimidiata (Linnaeus, 1758)	3, 11, 14, 21
Terebra felina (Dillwyn, 1817)	8, 18
Terebra funiculata Hinds, 1844	4, 8
Terebra guttata (Röding, 1798)	20, 22
Terebra jenningsi R.D. Burch, 1965	8, 28
Terebra laevigata Gray, 1834	2
Terebra maculata (Linnaeus, 1758)	3, 11, 12, 23
Terebra montgomeryi R.D. Burch, 1965	11, 21
Terebra nebulosa Sowerby, 1825	1, 4, 5, 10, 15, 18, 19
Terebra sp.	
Terebra solida Deshayes, 1857	4
Terebra subulata (Linnaeus, 1767)	3, 7, 19
Terebra succincta (Gmelin, 1791)	7, 14
Terebra triseriata Gray, 1834	
Terebra undulata Gray, 1834	3, 14, 23
Terenolla pygmaea (Hinds, 1844)	19
Family Conidae	
Conus aculeiformis Reeve, 1844	
Conus actangulus Lamarck, 1810	
Conus arenatus Hwass in Bruguière, 1792	1, 3, 5, 7, 11, 12, 15, 20-23, 26-28
Conus aulicus Linnaeus, 1758	17
Conus aurisiacus Linnaeus, 1758	22
Conus balteatus Sowerby, 1833	25
Conus boeticus Reeve, 1843	20
Conus canonicus Hwass in Bruguière, 1792	
Conus capitaneus Linnaeus, 1758	1, 4, 7, 12, 13
Conus catus Hwass in Bruguière, 1792	
Conus ceylanensis Hwass in Bruguière, 1792	4, 8
Conus chaldeus (Röding, 1798)	4, 13, 27
Conus circumcisus Born, 1778	
Conus cocceus Reeve, 1844	
Conus coronatus (Gmelin, 1791)	4, 8, 10, 12, 15, 17, 18, 20, 22, 25, 27
Conus distans Hwass in Bruguière, 1792	6, 25

continued

Species	Site records
Conus ebraeus Linnaeus, 1758	4, 13
Conus eburneus Hwass in Bruguière, 1792	14, 20
Conus emaciatus Reeve, 1849	7, 11, 22
Conus figulinus Linnaeus, 1758	
Conus flavidus Lamarck, 1810	4, 10
Conus frigidus Reeve, 1843	
Conus furvus Reeve, 1843	
Conus generalis Linnaeus, 1767	5
Conus geographus Linnaeus, 1758	
Conus glans Hwass in Bruguière, 1792	1, 9, 15
Conus imperialis Linnaeus, 1758	1, 6, 8, 10, 19, 21
Conus leopardus Röding, 1798	
Conus lischkeii Weinkauff, 1875	
Conus litoglyphus Röding, 1798	18
Conus litteratus Linnaeus, 1758	5, 7, 10, 12, 22
Conus lividus Hwass in Bruguière, 1792	10, 21, 27
Conus luteus Sowerby, 1833	4
Conus magus Linnaeus, 1758	6, 26, 27
Conus marmoreus Linnaeus, 1758	4, 15
Conus miles Linnaeus, 1758	1, 2, 4, 6, 7, 8, 12 ,16-18, 20, 24-28
Conus miliaris Hwass in Bruguière, 1792	8
Conus monachus Linnaeus, 1758	
Conus moreleti Crosse, 1858	1, 3, 4, 6, 9, 24, 25
Conus muriculatus Sowerby, 1833	12, 13
Conus musicus Hwass in Bruguière, 1792	2-4, 6-8, 10-13, 15, 16, 18-22, 27, 28
Conus mustelinus Hwass in Bruguière, 1792	22, 23, 25, 26, 28
Conus nussatella Linnaeus, 1758	3, 6, 7, 9, 22, 27
Conus omaria Hwass in Bruguière, 1792	
Conus parvulus Rlink, 1807	
Conus pertusus Hwass in Bruguière, 1792	27
Conus planorbis Born, 1778	19
Conus pulicarius Hwass in Bruguière, 1792	8, 10-12, 21, 22, 26, 27
Conus quercinus Lightfoot, 1786	14
Conus rattus Hwass in Bruguière, 1792	27
Conus retifer Menke, 1829	12
Conus sanguinolentus Quoy & Gaimard, 1834	
Conus scabriusculus Dillwyn, 1828	13
Conus sponsalis Hwass in Bruguière, 1792	4, 6, 8, 9, 10, 11, 12, 15, 19, 25, 27

Species	Site records
Conus stercmuscarum Linnaeus, 1758	27
Conus striatus Linnaeus, 1758	4, 8, 10
Conus suturatus Reeve, 1844	
Conus terebra Born, 1778	1, 3, 17, 21
Conus tessellatus Born, 1778	25
Conus textile Linnaeus, 1758	13, 19
Conus varius Linnaeus, 1758	8, 23, 27
Conus vexillum Gmelin, 1791	15, 19
Conus viola Cernohorsky, 1977	3, 9, 13
Conus virgo Linnaeus, 1758	7, 17, 19, 21, 25
Conus vitulinus Hwass in Bruguière, 1792	
Family Architectonicidae	
Heliacus variegata Gmelin, 1791	
Philippia radiata (Röding, 1798)	14, 19
Family Pyramidellidae	
Otopleura auriscati (Holten, 1802)	
Pyramidella acus (Gmelin, 1791)	
Pyramidella sulcata (Adams, 1852)	
Family Acteonidae	
Pupa alveola (Souverbie, 1863)	
Pupa solidula (Linnaeus, 1758)	3, 14
Pupa sulcata (Gmelin, 1791)	14
Family Hydatinidae	
Hydatina physis (Linnaeus, 1758)	
Family Cylichnidae	
Cylichna arachis (Quoy & Gaimard, 1833)	
Family Aglajidae	
Chelidonura electra Rudman, 1970	
Chelidonura fulvipunctata Baba, 1938	
Chelidonura hirundinina (Quoy & Gaimard, 1824)	5
Chelidonura inornata Baba, 1949	
Chelidonura sp. 1	
Chelidonura sp. 2	
Philinopsis gardineri (Eliot, 1903)	3, 7
Family Haminoeidae	
Atys cylindricus (Helbling, 1779)	3, 14
Atys naucum (Linnaeus, 1758)	7, 11, 12, 14, 23
Family Smaragdinellidae	
Smaragdinella sp.	27
Family Bullidae	
Bulla ampulla Linnaeus, 1758	14
Bulla vernicosa Gould, 1859	

continued

Species	Site records
Family Plakobranchidae	
Plakobranchus ocellatus van Hasselt, 1824	20, 22, 24, 25, 27
Family Elysiidae	
Elysia aff. *expansa* (O'Donoghue, 1924)	
Elysia sp.	
Elysia aff. *ornata* (Swainson, 1840)	
Thurdilla bayeri Marcus, 1965	6
Thurdilla sp. 1	16
Thurdilla sp. 2	17, 18
Family Caliphyllidae	
Cyerce sp.	
Family Aplysiidae	
Aplysia parvula Guilding, 1863	
Bursatella sp.	
Family Umbraculidae	
Umbraculum sinicum (Gmelin, 1783)	
Family Pleurobranchidae	
Pleurobranchus sp. 1	
Pleurobranchus sp. 2	
Family Polyceridae	
Nembrotha lineolata Bergh, 1905	
Nembrotha sp.	
Family Gymnodorididae	
Gymnodoris alba Baba, 1930	
Gymnodoris sp.	
Family Aegiritidae	
Notodoris minor Eliot, 1904	1, 2
Notodoris sp. *nov.*	
Family Hexabranchidae	
Hexabranchus sanguineus (Rüppell & Leuckart, 1828)	
Family Dorididae	
Ardedoris egretta Rudman, 1984	26
Discodoris boholensis Bergh, 1877	
Jorunna funebris (Kelaart, 1858)	25
Family Chromodorididae	
Chromodoris coi Risbec, 1956	
Chromodoris elisabethina Bergh, 1877	3
Chromodoris aff. *fidelis* (Kelaart, 1858)	
Chromodoris kuniei Pruvot-Fol, 1930	10
Chromodoris aff. *lineolata* Bergh, 1905	
Chromodoris lochi Rudman, 1982	5, 6, 9
Chromodoris magnifica Eliot, 1904	20

Species	Site records
Chromodoris cf. *striatella* Bergh, 1877	
Chromodoris tinctoria (Rüppell & Leuckart, 1828)	
Chromodoris willani Rudman, 1982	9
Chromodoris sp. 1	
Chromodoris sp. 2	
Glossodoris atromarginata (Cuvier, 1804)	10
Glossodoris aff. *rufomarginata* (Bergh, 1890)	
Family Phyllidiidae	
Fryeria menindie Brunckhorst, 1993	3
Phyllidia coelestis Bergh, 1905	1, 4, 5, 6, 7, 9, 18, 21, 26-28
Phyllidia elegans Bergh, 1869	6, 25
Phyllidia ocellata (Cuvier, 1804)	
Phyllidia pipeki Brunckhorst, 1993	25
Phyllidia varicosa Lamarck, 1801	4
Phyllidia sp.	
Phyllidiella pustulosa (Cuvier, 1804)	2-6, 9, 12, 15, 17, 24-28
Phyllidiella zelandica (Kelaart, 1859)	
Phyllidiopsis striata (Bergh, 1889)	
Reticulidia sp. 1	
Reticulidia sp. 2	
Family Flabellinidae	
Flabellina exoptata Gosliner & Willan, 1990	1
Family Aeolidiidae	
Godiva sp.	
Family Glaucidae	
Pteraeolidia ianthina (Angas, 1864)	
Family Onchidiidae	
Onchidium sp.	27
Family Siphonariidae	
Siphonaria atra (Quoy & Gaimard, 1833)	
Siphonaria denticulata (Quoy & Gaimard, 1833)	
Siphonaria javanica (Lamarck, 1819)	4
Siphonaria laciniosa Linnaeus, 1758	
Siphonaria sirius Pilsbry, 1894	10
Family Ellobiidae	
Cassidula nucleus (Gmelin, 1791)	
Ellobium sp.	
Ellobium aurisjudae Linnaeus, 1758	

continued

Species	Site records
Melampus fasciatus Deshayes, 1830	27
Pythia scabraeus (Linnaeus, 1758)	4
CLASS BIVALVIA	
Family Mytilidae	
Amygdalum politum (Verrill & Smith, 1880)	14
Brachidontes sp.	
Lithophaga sp.	1, 3, 4, 6-11, 16, 19-23, 25, 26, 28
Lithophaga sp. 2	
Modiolus philippinarum Hanley, 1843	22, 23, 26
Modiolus sp.	18
Septifer bilocularis (Linnaeus, 1758)	10, 15, 21, 27
Xenostrobus sp. 1	
Xenostrobus sp. 2	
Family Arcidae	
Anadara antiquata (Linnaeus, 1758)	23
Anadara granosa (Linnaeus, 1758)	
Anadara maculosa (Reeve, 1844)	23, 26
Arca avellana (Lamarck, 1819)	1, 3, 7, 9, 11, 12, 15, 17-27
Arca navicularis Bruguière, 1798	
Arca ventricosa (Lamarck, 1819)	11, 23, 24, 28
Barbatia amygdalumtotsum (Röding, 1798)	1, 2, 5, 7-12, 18, 20, 21, 24
Barbatia foliata Forskål, 1775	1, 17
Barbatia pistachia (Lamarck, 1819)	5
Barbatia plicata (Dillwyn,m 1817)	1, 8, 10, 12, 13
Barbatia tenella (Reeve, 1843)	3
Barbatia trapezina (Lamarck, 1819)	13
Barbatia ventricosa (Lamarck, 1819)	4, 5
Trisidos semitorta (Lamarck, 1819)	
Family Glycymerididae	
Glycymeris reevei (Mayer, 1868)	1, 3, 5, 19, 27, 28
Tucetona amboiensis (Gmelin, 1791)	
Tucetona petunculus (Linnaeus, 1758)	5, 8, 13
Family Pteriidae	
Pinctada margaritifera (Linnaeus, 1758)	4-6, 8-11
Pinctada maxima (Jameson, 1901)	12
Pteria avicular (Holten, 1802)	8, 9, 12, 17
Pteria pengiun (Röding, 1798)	4-6, 12, 18
Family Malleidae	
Malleus malleus (Linnaeus, 1758)	
Plicatula chinensis Mörch, 1853	11
Vulsella vulsella (Linnaeus, 1758)	12

Species	Site records
Family Isognomonidae	
Isognomon perna (Linnaeus, 1767)	12, 13
Isognomon sp.	
Family Pinnidae	
Atrina vexillum (Born, 1778)	4
Pinna bicolor (Gmelin, 1791)	6, 10, 14, 20, 23
Pinna muricata (Linnaeus, 1758	
Streptopinna saccata (Linnaeus, 1758)	5, 7, 9, 11, 28
Family Limidae	
Ctenoides ales (Finlay, 1927)	
Ctenoides annulata (Lamarck, 1819)	1, 4, 5, 20, 22, 24
Lima cf. *basilanica* (A. Adams & Reeve, 1850)	15
Lima fragilis (Gmelin, 1791)	2
Lima lima (Link, 1807)	1, 5, 6, 9, 12, 15, 17, 20, 25
Family Ostreidae	
Alectryonella plicatula (Gmelin, 1791)	
Alectryonella sp.	
Dendostrea folium (Linnaeus, 1758)	11
Dendostrea sandwichensis (Sowerby in Reeve, 1871)	11
Hyotissa hyotis (Linnaeus, 1758)	2, 4-6, 11, 23, 26
Lopha cristagalli (Linnaeus, 1758)	4, 11, 18, 23, 26
Lopha sp. 2	
Parahyotissa imbricata (Lamarck, 1819)	1, 2, 5
Saccostrea cf. *cucullata* (Born, 1778)	8, 11, 25-27
Saccostrea echinata (Quoy & Gaimard, 1835)	
Saccostrea sp. 1	
Saccostrea sp. 2	
Family Plicatulidae	
Plicatula cf. *muricata* Sowerby, 1873	11
Family Pectinidae	
Anguipecten cf. *aurantiacus* (Röding, 1798)	25
Annachlamys reevei (Adams & Reeve, 1850)	1
Chlamys corsucans (Hinds, 1845)	
Chlamys irregularis (Sowerby, 1842)	7
Chlamys lentiginosa (Reeve, 1865)	10, 20
Chlamys madreporarum (Sowerby, 1842)	
Chlamys mollita (Reeve, 1853)	1
Chlamys squamosa (Gmelin, 1791)	4, 8, 20
Comptopallium radula (Linnaeus, 1758)	1, 4, 7-10, 15, 21, 22

continued

Species	Site records
Exichlamys histronica (Gmelin, 1791)	10
Exichlamys spectabilis (Reeve, 1853)	24
Glorichlamys elegantissima Deshayes, 1863	5
Gloripallium pallium (Linnaeus, 1758)	1, 3, 7, 21, 22, 24, 26-28
Gloripallium speciosa (Reeve, 1853)	2
Laevichamys brettinghami Dijkstra, 1998	1, 9
Mimachlamys lentiginosa (Reeve, 1853)	
Mimachlamys sp.	
Mirapecten moluccensis Dijkstra, 1988	24
Mirapecten rastellum (Lamarck, 1819)	
Pedum spondyloidaeum (Gmelin, 1791)	1-6, 9, 11, 12, 15-28
Semipallium fulvicostatum (Adams & Reeve, 1850)	3, 10
Semipallium tigris (Lamarck, 1819)	3-5, 12, 17, 21
Family Spondylidae	
Spondylus candidus (Lamarck, 1819)	1-6, 10, 16, 21, 23, 24, 26-28
Spondylus multimuricatus Reeve, 1856	
Spondylus prionifer Iredale, 1931	11
Spondylus sanguineus Dunker, 1852	1, 7, 12, 13, 26
Spondylus sinensis Schreibers, 1793	3, 7, 8, 20
Spondylus squamosus Schreibers, 1793	
Spondylus varians Sowerby, 1829	10
Family Anomiidae	
Anomia sp.	1
Family Placunidae	
Placuna lobata Sowerby, 1871	
Family Chamidae	
Chama cf. *asperella* Lamarck, 1819	13
Chama brassica Reeve, 1846	5
Chama fibula Reeve, 1846	
Chama lazarus Linnaeus, 1758	
Chama limbula (Lamarck, 1819)	4, 10, 18, 19, 21, 23-26, 28
Chama pacifica Broderip, 1834	5
Chama plinthota Cox, 1927	10
Chama savigni Lamy, 1921	
Chama sp.	
Family Lucinidae	
Anodontia sp.	1
Anodontia edentula (Linnaeus, 1758)	
Anodontia pila (Reeve, 1850)	
Ctena bella (Conrad, 1834)	3
Divaricella ornata (Reeve, 1850)	3

Species	Site records
Family Fimbriidae	
Codakia punctata (Linnaeus, 1758)	21
Codakia paytenorum (Iredale, 1937)	14
Codakia tigerina (Linnaeus, 1758)	13, 21, 22, 27
Fimbria fimbriata (Linnaeus, 1758)	11, 28
Family Galeommatidae	
Scintilla sp	
Family Carditidae	
Beguina semiorbiculata (Linnaeus, 1758)	
Cardita abyssicola Hinds, 1843	14
Cardita variegata Bruguière, 1792	2, 4, 20
Megacardita aff *incrassata* (Sowerby, 1825)	
Family Cardiidae	
Acrosterigma alternatum (Sowerby, 1841)	
Acrosterigma angulata (Lamarck, 1819)	1, 4, 7, 11, 15, 20, 21
Acrosterigma elongata (Bruguière, 1789)	
Acrosterigma flava (Linnaeus, 1758)	
Acrosterigma fovealatum (Sowerby, 1840)	10, 18
Acrosterigma impolita (Sowerby, 1833)	3
Acrosterigma mendanaense (Sowerby, 1896)	1-6, 8, 12, 16-22, 28
Acrosterigma reeveanum (Dunker, 1852)	
Acrosterigma transcendens (Melvill & Standen, 1899)	1-3, 8, 10
Acrosterigma unicolor (Sowerby, 1834)	1, 2, 5, 7, 21, 22
Acrosterigma cf. *vlamingi* Wilson & Stevenson, 1978	
Corculum dionaeum (Broderip & Sowerby, 1829)	
Ctenocardia fornicata (Sowerby, 1840)	2
Fragum fragum (Linnaeus, 1758)	3, 10, 13, 22
Fragum retusum (Linnaeus, 1767)	10
Fragum unedo (Linnaeus, 1758)	10,13,14,21
Fulvia aperta (Bruguière, 1789)	
Vepricardium multispinosum (Sowerby, 1838)	14
Family Tridacnidae	
Hippopus hippopus (Linnaeus, 1758)	4, 7, 10-12, 21
Tridacnea crocea Lamarck, 1819	1, 3, 4, 7, 11-13, 15, 18, 20, 21, 23-28
Tridacna derasa (Röding, 1798)	7, 11
Tridacna gigas (Linnaeus, 1758)	11, 13, 15, 17, 20, 23
Tridacna maxima (Röding, 1798)	11-13

continued

Species	Site records
Tridacna squamosa Lamarck, 1819	1, 3-7, 11-13, 15, 17, 18, 20, 21, 22, 25-28
Family Mactridae	
Mactra abbreviata Lamarck, 1819	
Mactra alta Deshayes, 1854	
Mactra cf. *eximia* Reeve, 1854	14
Mactra sp.	
Spisula cf. *aspersa* (Sowerby, 1825)	21
Family Tellinidae	
Exotica assimilis (Hanley, 1844)	19
Macoma cf. *consociata* (Smith, 1885)	23
Tellina capsoides Lamarck, 1818	
Tellina exculta Gould, 1850	21
Tellina inflata Gmelin, 1791	27
Tellina gargadia Linnaeus, 1758	3, 7,10
Tellina linguafelis Linnaeus, 1758	11, 20, 21, 24, 26
Tellina ovalis Sowerby, 1825	3
Tellina palatum (Iredale, 1929)	2, 3, 7, 11, 12, 21, 28
Tellina perna Splengler, 1798	
Tellina phaoronis Hanley, 1844	14
Tellina plicata Valenciennes, 1827	3, 14
Tellina pretium Salisbury, 1934	
Tellina rastellum Hanley, 1844	3
Tellina remies Linnaeus, 1758	21
Tellina rostrata Linnaeus, 1758	3, 7
Tellina serricostata Tokunaga, 1906	28
Tellina scobinata Linnaeus, 1758	4, 19, 20, 26
Tellina staurella Lamarck, 1818	2, 3, 7, 22, 23
Tellina cf. *tenuilamellata* Smith, 1885	4
Tellina tongana Quoy & Gaimard, 1835	10
Tellina virgata Linnaeus, 1758	3, 4
Family Semelidae	
Semele cf. *australis* (Sowerby, 1832)	15
Semele lamellosa (Sowerby, 1830)	28
Family Psammobiidae	
Asaphis violaceans (Forskål, 1775)	1, 13, 26
Gari amethystus (Wood, 1815)	16
Gari maculosa (Lamarck, 1818)	12
Gari occidens (Gmelin, 1791)	
Gari pennata (Deshayes, 1855)	1, 18, 19
Gari rasilis (Melvill and Standen, 1899)	
Gari squamosa (Lamarck, 1818)	3, 4, 7, 10, 11
Family Solecurtidae	
Solecurtis philippinarum (Dunker, 1861)	
Solecurtis sulcata (Dunker, 1861)	

Species	Site records
Family Donacidae	
Donax faba Gmelin, 1791	10
Donax sp.	
Family Trapeziidae	
Trapezium bicarinatum (Schumacher, 1817)	
Trapezium obesa (Reeve, 1843)	2
Family Veneridae	
Antigona clathrata (Deshayes, 1854)	12
Antigona chemnitzii (Hanley, 1844)	18, 19, 22
Antigona persimilis (Iredale, 1930)	
Antigona purpurea (Linnaeus, 1771)	
Antigona restriculata (Sowerby, 1853)	2, 3, 20, 28
Antigona reticulata (Linnaeus, 1758)	16, 17, 19
Bassina sp.	22
Callista impar (Lamarck, 1818)	
Callista lilacina (Lamarck, 1818)	3, 14
Callista semisulcata (Sowerby, 1851)	14
Circe lenticularis (Deshayes, 1853)	
Circe scripta (Linnaeus, 1758)	11
Circe sulcaa Gray, 1838	
Dosinia amphidesmoides (Reeve, 1850)	
Dosinia incisa (Reeve, 1850)	
Dosinia iwakawai Oyama & Habe, 1970	14, 16, 26
Dosinia juvenilis (Gmelin, 1791)	14
Dosinia aff. *lucinalis* (Lamarck, 1835)	
Dosinia mira Smith, 1885	3
Dosinia cf. *tumida* (Gray, 1838)	
Dosinia sculpta (Hanley, 1845)	1
Gafrarium aequivocum (Holten, 1803)	12, 23
Gafrarium menkei (Jonas, 1846)	
Gafrarium pectinatum (Holten, 1802)	
Gafrarium tumidum Röding, 1798	4
Globivenus capricornsue (Hedley, 1908)	
Globivenus toreuma (Gould, 1850)	1, 2, 4, 6-9, 12, 16-20
Gomphina sp.	1
Lioconcha annettae Lamprell & Whitehead, 1990	1, 3, 4, 7, 8, 11
Lioconcha castrensis (Linnaeus, 1758)	4, 5, 8, 10, 12, 13, 19-22, 24, 26-28
Lioconcha fastigiata (Sowerby, 1851)	11, 12
Lioconcha ornata (Dillwyn, 1817)	1, 20, 23
Lioconcha polita (Röding, 1798)	12, 14, 21
Paphia sp.	3, 19, 20

continued

Species	Site records
Pitar affinis (Gmelin, 1791)	11, 14
Pitar cf. *nancyae* Lamprell & Whitehead, 1990	27
Pitar prora (Conrad, 1837)	
Pitar spoori Lamprell & Whitehead, 1990	12, 26
Pitar subpellucidus (Sowerby, 1851)	
Samarangia quadrangularis (Adams & Reeve, 1850)	
Tapes dorsatus (Lamarck, 1818)	
Tapes literatus (Linnaeus, 1758)	21
Tapes platyptycha Pilsbry, 1901	
Tapes sulcarius Lamarck, 1818	1, 4
Tawera torresiana (Smith, 1884)	4, 19, 22
Timoclea costillifera (Adams & Reeve, 1850)	10
Timoclea marica (Linnaeus, 1758)	
Family Corbulidae	
Corbula sp.	10, 11
Corbula macgillvrayi Smith, 1885	10, 23
Corbula cf *taheitensis* Lamarck, 1818	
Polymesoda coaxans (Gmelin, 1791)	
CLASS CEPHALOPODA	
Family Nautilidae	
Nautilus pompilius Linnaeus, 1758	
Family Spirulidae	
Spirula spirula (Linnaeus, 1758)	
Family Sepiidae	
Sepia sp.	13
Family Argonautidae	
Argonauta argo Linnaeus, 1758	
Family Octopodidae	
Octopus sp.	8
CLASS SCAPHOPODA	
Family Dentaliidae	
Dentalium aprinum Linnaeus, 1766	3
Dentalium crocinum (Dall, 1907)	
Dentalium sp.	14
Dentalium elephantinum Linnaeus, 1758	

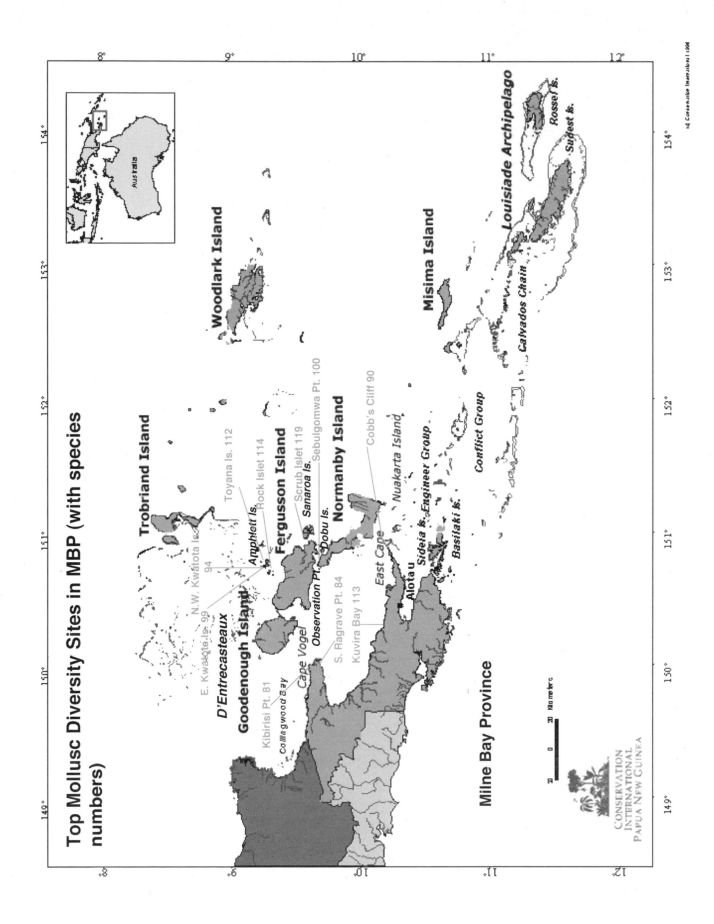

Top Mollusc Diversity Sites in MBP (with species numbers)

Appendix 5

List of the reef fishes of Milne Bay Province, Papua New Guinea

Gerald R. Allen

This list includes all species of shallow (to 60 m depth) coral reef fishes known from Milne Bay Province at 1 July 2000. The list is based on the following sources:

1) Results of the 1997 CI Marine RAP;

2) a cumulative list provided by Bob Halstead based chiefly on observations and collecting activities by himself, J.E.Randall (Bishop Museum, Hawaii), and Rudie Kuiter (Museum of Victoria);

3) a 10 day visit to MBP by G. Allen in November 1999, and

4) results of the CI Marine RAP between 30 May and 26 June 2000. The family classification follows that of Eschmeyer (1998) except for the placement of Cirrhitidae.

Terms relating to relative abundance are as follows:

Abundant - Common at most sites in a variety of habitats with up to several hundred individuals being routinely observed on each dive.

Common - seen at the majority of sites in numbers that are relatively high in relation to other members of a particular family, especially if a large family is involved.

Moderately common - not necessarily seen on most dives, but may be relatively common when the correct habitat conditions are encountered.

Occasional - infrequently sighted and usually in small numbers, but may be relatively common in a very limited habitat.

Rare - less than 10, often only one or two individuals seen on all dives.

Species indicated wth an asterisk represent additions to the list published by Allen (1997). The numbers under the site records column and remarks in the abundance column pertain to the 2000 survey.

Summary of fish diversity sites are presented in the map following the table.

Species	Site records	Abundance	Depth (m)
ORECTOLOBIDAE			
Eucrossorhinus dasypogon (Bleeker, 1867)*	37	Rare, a single specimen recorded.	
HEMISCYLLIIDAE			
Chiloscyllium punctatum Müller and Henle, 1838	Seen on 1997 survey only.		1-10
Hemiscyllium hallstromi Whitley, 1967	Recorded previously by B. Halstead.		2-15
H. trispeculare Richardson, 1843	13,17	Rare, only two individuals seen.	1-15
GINGLYMOSTOMATIDAE			
Nebrius ferrugineus (Lesson, 1830)*	30	Rare, a single individual recorded.	1-70
CARCHARHINIDAE			
Carcharhinus albimarginatus (Rüppell, 1837)	45	Rare, only one seen during survey.	14-40
C. amblyrhynchos (Bleeker, 1856)	6, 9, 16, 19, 30, 35, 37, 38, 44, 45, 47, 55, 57	Occasional, infrequently sighted during survey, except common on outer reef at Rossel, where approximately 15 seen at sites 44-45.	0-100
C. melanopterus (Quoy and Gaimard, 1824)	31, 36, 43	Rare, less than five individuals observed.	0-10
Galeocerdo cuvier (Péron & Lesueur, 1822)*	9	Rare, only one individual (about 200 cm TL) seen.	0-150
Triaenodon obesus (Rüppell, 1835)_	1, 6, 9, 12, 16, 31, 35, 40, 43-45, 52, 53	Occasional, usually on outer slopes. Photographed.	2-100
SPHYRNIDAE			
Sphyrna lewini (Griffith and Smith, 1834)	1	Rare, a single individual observed.	0-275
DASYATIDIDAE			
Dasyatis kuhlii (Müller and Henle, 1841)	32, 35, 49, 54, 56	Occasionally seen in sandy areas.	2-50
Pastinachus sephen (Forsskål, 1775)	Seen on 1997 survey only.		2-60
Taeniura lymma (Forsskål, 1775)	1, 4, 7, 14, 23, 25, 26, 28-35, 39, 40, 43, 45, 49, 50, 54	Occasionally seen in sandy areas.	2-30
Taeniura meyeni (Müller and Henle, 1841)*	31	Rare, a single individual observed.	1-200
Urogymnus asperrimus (Bloch and Schneider, 1801)	Seen on 1997 survey only.		2-100
MYLIOBATIDAE			
Aetobatus narinari (Euphrasen, 1790)	Recorded previously by B. Halstead.		0-25
MOBULIDAE			
Manta birostris (Walbaum, 1792)	1	Rare, only one individual observed.	0-100
MORINGUIDAE			
Moringua javanica (Kaup, 1856)	Seen on 1997 survey only.		1-10
M. microchir Bleeker, 1853	Seen on 1997 survey only.		3-20

continued

Species	Site records	Abundance	Depth (m)
CHLOPSIDAE			
Chilorhinus platyrhynchus (Norman, 1922)	Recorded previously by B. Halstead.		5-25
Kaupichthys hypoproroides (Strömann, 1896)	37	One specimen collected with rotenone	5-25
MURAENIDAE			
Echidna nebulosa (Thünberg, 1789)	Seen on 1997 survey only.		1-10
Enchelycore bayeri (Schultz, 1952)	Recorded previously by B. Halstead.		0-20
Gymnothorax bredeni McCosker and Randall, 1977	Seen on 1997 survey only.		5-40
G. buroensis (Bleeker, 1857)	Seen on 1997 survey only.		1-25
G. chilospilus Bleeker, 1865	Seen on 1997 survey only.		1-45
G. favagineus (Bloch & Schneider, 1801)	Recorded previously by B. Halstead.		1-40
G. fimbriatus (Bennett, 1831)	Seen on 1997 survey only.		0-30
G. flavimarginatus (Rüppell, 1828)	Seen on 1997 survey only.		1-150
G. javanicus (Bleeker, 1865)	6, 9, 11, 16, 19, 28, 30, 39, 41, 43, 44, 47, 48, 50, 54-57	Occasionl, the most common moray observed during RAP. Photographed.	0.5-50
G. melatremus Schultz, 1953	45	A single specimen collected with rotenone.	5-30
G. meleagris (Shaw and Nodder, 1795)	Seen on 1997 survey only.		1-40
G. zonipectus Seale, 1906	53	A single specimen collected with rotenone.	8-45
Rhinomuraena quaesita Garman, 1888	Seen on 1997 survey only.		1-50
Siderea thrysoidea (Richardson, 1845)	Recorded previously by B. Halstead.		0-8
Uropterygius marmoratus (Lacepède, 1803)	Seen on 1997 survey only.		1-20
U. sp.	Recorded previously by B. Halstead.		5-15
OPHICHTHIDAE			
Brachysomophis sp.	Recorded previously by B. Halstead.		1-15
Callechelys marmoratus (Bleeker, 1852)	Seen on 1997 survey only.		1-15
Leiuranus semicinctus (Lay and Bennett, 1839)	Seen on 1997 survey only.		0-20
Myrichthys colubrinus (Boddaert, 1781)	Recorded previously by B. Halstead.		0-8
Ophichthus bonaparti (Kaup, 1856)	Recorded previously by B. Halstead.		5-25
Shultzidia johnstonensis (Schultz & Woods, 1949)*	37	One specimen collected with rotenone.	5-30 m
CONGRIDAE			
Gorgasia maculata Klausewitz & Eibesfeldt, 1959	14, 19	Occasional, but locally common.	20-50
G. preclara Bohlke & Randall, 1981	Recorded previously by B. Halstead.		25-40

continued

Species	Site records	Abundance	Depth (m)
Heteroconger hassi (Klausewitz and Eibl-Eibesfeldt, 1959)	31, 39, 49, 50	Occasional colonies on sandy slopes.	3-45
H. polyzona (Bleeker, 1878)	Recorded previously by B. Halstead.		1-6
H. taylori Castle and Randall, 1995	Seen on 1997 survey only.		8-20
CLUPEIDAE			
Amblygaster sirm (Walbaum, 1792)	41	A school of about 50 fish seen.	0-3
Herklotsichthys quadrimaculatus (Rüppell, 1837)	Recorded previously by B. Halstead.		0-3
Sardinella melanura (Cuvier, 1829)	Recorded by GRA in 1999		0-3
Spratelloides delicatulus (Bennett, 1832)	3, 9-13, 15, 17, 22, 34, 42, 43, 45	Occasional, hundreds seen schooling near surface at several sites.	0-1
ENGRAULIDAE			
Thryssa sp.	Recorded previously by B. Halstead.		0-4
PLOTOSIDAE			
Plotosus lineatus (Thünberg, 1787)	8-10, 35, 40, 56	Occasional, several schools of juveniles containing up to about 200 fishes observed. Photographed.	1-20
SYNODONTIDAE			
Saurida gracilis (Quoy & Gaimard, 1824)	20, 26, 32, 38, 42, 46, 51	Occasional on sand bottoms.	1-30
Synodus binotatus Schultz, 1953	48	Rare, only one seen.	1-30
S. dermatogenys Fowler, 1912	3, 5, 11-13, 20, 26, 27, 29, 37-39, 42, 43, 48, 53, 56	Moderately common, solitary individuals usually seen resting on dead coral or rubble. Photographed.	1-25
S. jaculum Russell and Cressy, 1979	29, 31, 32, 45, 58	Occasional on rubble bottoms in the Louisiades.	10-50
S. rubromarmoratus Russell and Cressy, 1979	10	Rare, on sand or rubble bottoms. Photographed.	5-30
S. variegatus (Lacepède, 1803)	4, 5, 10, 24, 26, 30, 32, 34, 35, 37, 39, 42-46, 48-51, 54, 57	Moderately common, solitary individuals or pairs usually seen resting on live coral. Photoraphed.	5-50
Trachinocephalus myops (Bloch & Schneider, 1801)	Recorded previously by B. Halstead.		1-400
OPHIDIIDAE			
Brotula multibarbata (Temminck and Schlegel, 1846)	Seen on 1997 survey only.		5-150
CARAPIDAE			
Encheliophis homei (Richardson, 1844)	Recorded previously by B. Halstead.		2-30
BYTHITIDAE			
Brosmophyciops pautzkei Schultz, 1960	Seen on 1997 survey only.		5-55
Ogilbia sp.	53	Collected with rotenone.	0-5

continued

Species	Site records	Abundance	Depth (m)
ANTENNARIIDAE			
Antennarius commersonii (Latreille, 1804)	Recorded previously by B. Halstead.		1-40
A. maculatus (Desjardins, 1840)	Recorded previously by B. Halstead.		1-15
A. pictus (Shaw & Nodder, 1794)	Recorded previously by B. Halstead.		1-15
A. striatus (Shaw, 1794)	Recorded previously by B. Halstead.		10-200
GOBIESOCIDAE			
Diademichthys lineatus (Sauvage, 1883)	11, 13, 32, 36, 51, 57	Occasional among sea urchins or branching coral. Photographed.	3-20
ATHERINIDAE			
Atherinomorus lacunosus (Forster, 1801)	7	One specimen netted at surface.	0-2
Hypoatherina valenciennesi (Bleeker, 1853)*	34	Locally abundant at one site. Collected.	0-2
Stenatherina panatela (Jordan & Richardson, 1908)*	34, 41, 42, 43	Locally abundant at four sites. Collected and photographed.	0-4
BELONIDAE			
Strongylura incisa (Valenciennes, 1846)	Recorded previously by B. Halstead.		0-2
S. leiura (Bleeker, 1850)	Recorded previously by B. Halstead.		0-2
Tylosurus crocodilus (Peron & Lesueur, 1821)	10, 14, 26, 27, 29, 35, 37, 39, 45	Occasional, on surfaces at several sites.	0-4
HEMIRAMPHIDAE			
Hemirhamphus far (Forsskål, 1775)	34	Several seen at surface.	0-2
Hyporhamphus affinis (Günther, 1866)	2, 10, 34, 54	Three schools seen at surface.	0-2
H. dussumieri (Valenciennes, 1846)	Recorded previously by B. Halstead.		0-2
Zenarchopterus dispar (Valenciennes, 1847)	24	Generally rare, but usually seen on mangrove shores.	0-2
Discotrema crinophila Briggs, 1976	Seen on 1997 survey only.		5-30
ANOMALOPIDAE			
Anomalops katoptron (Bleeker, 1856)	Seen on 1997 survey only.		15-400
Photoblepharon palpebratus (Boddaert, 1781)	Recorded previously by B. Halstead.		10-100
HOLOCENTRIDAE			
Myripristis adusta Bleeker, 1853	2, 4, 6, 16, 22, 28, 30, 35, 39, 43-45, 48, 49, 51, 54		3-30
M. berndti Jordan and Evermann, 1902	4-6, 9, 19, 22, 28-31, 37, 39, 40, 43-45, 49, 50, 52-55	Occasional, sheltering in caves and under ledges. Common at site 45.	8-55
M. hexagona (Lacepède, 1802)	14, 26-28, 38, 51, 57	Occasional, usually in coastal areas affected by silt.	10-40

continued

Species	Site records	Abundance	Depth (m)
M. kuntee Valenciennes, 1831	6, 9, 17, 18, 26, 29-35, 37-40, 43-46, 48-51, 53-56	Moderately common, sheltering in caves and under ledges, but frequently exposes itself for brief periods. Very common at site 55.	5-30
M. murdjan (Forsskål, 1775)	1, 18, 19, 31, 32, 37, 45, 49, 55-57	Moderately common, sheltering in caves and under ledges.	3-40
M. pralinia Cuvier, 1829	5, 14, 24, 26-28, 38, 51, 57	Moderately common, but shleters deep in crevices during the day. Photographed.	3-40
M. violacea Bleeker, 1851	1, 3-9, 11-13, 15-18, 20-24, 26-33, 35, 37-46, 48-51, 53-57	Common, most abundant squirrelfish seen in MBP. Often seen at entrance of crevices. Most abundant at site 45.	3-30
M. vittata Valenciennes, 1831	2, 5, 18, 19, 22, 24, 28, 30, 37, 39, 43, 44, 52, 53	Moderately common, sheltering in caves and ledges on drop-offs. Especially common at sites 44, 52, and 53. Photographed.	12-80
Neoniphon argenteus (Valenciennes, 1831)	Seen on 1997 survey only.		3-30
N. aurolineatus (Valenciennes)	30	Three seen in 40 m depth.	30-160
N. opercularis (Valenciennes, 1831)	11, 19, 26, 36, 39, 46, 51, 54	Occasional.	3-20
N. sammara (Forsskål, 1775)	1, 3, 4, 7, 14, 16-19, 21, 24, 26-32, 34-36, 38-40, 42-57	Moderately common, usually among branches of staghorn Acropora coral. Especially abundant at sites 42 and 55. Photographed.	2-50
Plectrypops lima (Valenciennes, 1831)	Seen on 1997 survey only.		15-50
Sargocentron caudimaculatum (Rüppell, 1835)	2, 4-6, 9, 17-20, 28-31, 34, 39, 42-45, 47-50, 52-55	Moderately common, always seen close to cover. Photographed.	6-45
S. cornutum (Bleeker, 1853)	15, 16, 36, 51	Occasional. Photographed.	6-50
S. diadema (Lacepède, 1802)	29-32, 34, 39, 42, 44, 45, 48-52, 55	Moderately common. Especially abundant at sites 30 and 55. Photographed.	2-30
S. ittodai (Jordan and Fowler, 1903)	Seen on 1997 survey only.		6-70
S. melanospilos (Bleeker, 1858)	36, 51	Rare, only a few seen at two sites.	10-25
S. microstomus (Günther, 1859)	43, 44, 52	Only five fish seen, but this nocturnal species is probably not uncommon.	1-180
S. praslin (Lacepède, 1802)	Seen on 1997 survey only.		2-15
S. rubrum (Forsskål, 1775)*	11, 56	Rare, only two seen.	5-25
S. spiniferum (Forsskål, 1775)	6-8, 10, 11, 13, 18, 21, 28-31, 33, 34, 36, 39, 41, 42, 46-48, 50, 51, 54-57	Moderately common, in caves and under ledges. Photographed.	5-122
S. tiere (Cuvier, 1829)	52, 53	Rarely seen, but is nocturnal.	10-180

continued

Species	Site records	Abundance	Depth (m)
S. tieroides (Bleeker, 1853)	5, 38	Rarely seen, but is nocturnal.	10-40
S. violaceus (Bleeker, 1853)	9, 17, 31, 35, 51, 54	Occasional.	3-30
PEGASIDAE			
Eurypegasus draconis (Linnaeus, 1766)	Recorded previously by B. Halstead.		2-20
AULOSTOMIDAE			
Aulostomus chinensis (Linnaeus, 1766)	3, 4, 12, 31, 35, 39, 43, 48, 54	Occasional.	2-122
FISTULARIIDAE			
Fistularia commersoni Rüppell, 1835	2, 9, 14, 16, 19, 32, 39, 54	Occasional.	2-128
CENTRISCIDAE			
Aeoliscus strigatus (Günther, 1860)	12, 14, 26, 57.	Occasional schools observed.	1-30
Centriscus scutatus (Linnaeus, 1758)	Recorded previously by B. Halstead.		1-30
SOLENOSTOMIDAE			
Solenostomus armatus Weber, 1913	Recorded previously by B. Halstead.		5-25
S. cyanopterus Bleeker, 1854	Recorded previously by B. Halstead.		5-25
S. paradoxus (Pallas, 1770)	Recorded previously by B. Halstead.		5-25
SYNGNATHIDAE			
Acentronura tentaculata Günther, 1870	Recorded previously by B. Halstead.		2-15
Corythoichthy amplexus Dawson & Randall, 1975	Recorded previously by B. Halstead.		8-25
Corythoichthys flavofasciatus (Rüppell, 1838)	Seen on 1997 survey only.	Moderately common at several sites exposed to cool temperatures.	1-25
C. haematopterus (Bleeker, 1851)	Recorded previously by B. Halstead.		1-20
C. intestinalis (Ramsay, 1881)	35, 26, 47	Only seen at three sites and in low numbers.	1-25
C. nigripectus Herald, 1953	Recorded previously by B. Halstead.		4-30
C. ocellatus Herald, 1953	Seen on 1997 survey only.		1-15
C. schultzi Herald, 1953	Seen on 1997 survey only.		1-30
Doryrhamphus dactyliophorus (Bleeker, 1853)	20, 53	Only two seen, but a secretive cave and ledge dweller.	1-56
D. excisus Kaup, 1856	Seen on 1997 survey only.		2-50
D. janssi (Herald & Randall, 1972)	Recorded previously by B. Halstead.		5-35
Halicampus macrorhynchus Bamber, 1915	Recorded previously by B. Halstead.		3-30
Halicampus nitidus (Günther, 1873)	Seen on 1997 survey only.	One specimen collected with rotenone.	3-25
Hippocampus bargibanti Whitley, 1970	Seen by GRA in 1999		10-40
H. kuda Bleeker, 1852	Recorded previously by B. Halstead.		0-12

continued

Species	Site records	Abundance	Depth (m)
H. sp.	Recorded previously by B. Halstead.		1-15
Siokunichthys nigrolineatus Dawson, 1983	12, 29	Rare, only a few seen. Commensal with mushroom corals.	10-20
Syngnathoides biaculeatus (Bloch, 1785)	Recorded previously by B. Halstead.		0-10
Trachyramphus bicoarctata (Bleeker, 1857)	Recorded previously by B. Halstead.		1-42
T. longirostris Kaup, 1856	Recorded previously by B. Halstead.		10-90
SCORPAENIDAE			
Dendrochirus biocellatus (Fowler, 1935)	Recorded previously by B. Halstead.	One specimen seen in cave.	1-40
D. brachypterus (Cuvier, 1829)	Recorded previously by B. Halstead.		1-15
D. zebra (Cuvier, 1829)	36, 42	Rarely seen, but most active at night.	1-20
Parascorpaena mcadamsi (Fowler, 1938)	45	Collected with rotenone in 18 m depth.	3-20
Parascorpaena mossambica (Peters, 1855)	Recorded previously by B. Halstead.		1-18
Pteroidichthys amboinensis Bleeker, 1856	Recorded previously by B. Halstead.		3-20
Pterois antennata (Bloch, 1787)	2-5, 9, 13, 19, 26, 28, 37, 48	Occasional, but mainly nocturnal. Photographed.	1-50
P. volitans (Linnaeus, 1758)	2, 4, 5, 9, 19, 22, 31, 34, 36, 39, 42	The most commonly seen scorpionfish, but only occasional sightings.	2-50
Rhinopias aphanes Eschmeyer, 1973	Recorded by GRA in 1999.		10-50
R. frondosa (Günther, 1891)	Recorded previously by B. Halstead.		13-90
Scorpaenodes albaiensis Evermann and Seale, 1907	53	One specimens collected with rotenone.	8-40
S. guamensis (Quoy and Gaimard, 1824)	Seen on 1997 survey only.		0-10
S. hirsutus (Smith, 1957)	Seen on 1997 survey only.		5-40
S. parvipinnis (Garrett, 1863)	Seen on 1997 survey only.		2-50
S. varipinnis Smith, 1957	Seen on 1997 survey only.		1-50
Scorpaenopsis diabolus (Cuvier, 1829)	Seen on 1997 survey only.		1-70
S. oxycephala (Bleeker, 1849)	Seen on 1997 survey only.	Rare, only one seen.	1-40
S. macrochir Ogilby, 1910	23	Rare, only one seen, but diffult to detect. Photographed.	1-10
S. venosus (Cuvier, 1829)	Seen on 1997 survey only.		3-40
Sebastapistes cyanostigma (Bleeker, 1856)	18	Probably not uncommon, but only one seen among coral branches.	2-15
S. strongia (Cuvier, 1829)	Seen on 1997 survey only.		1-15
Taenianotus triacanthus Lacepède, 1802	Seen on 1997 survey only.	Rare, a single individual observed.	5-130

continued

continued

Species	Site records	Abundance	Depth (m)
TETRARODIGAE			
Ablabys taenianotus (Cuvier, 1829)	Recorded previously by B. Halstead.		1-10
SYNANCEIIDAE			
Erosa erosa (Langsdorf, 1829)	Recorded previously by B. Halstead.		1-10
Inimicus didactylus (Pallas, 1769)	Recorded previously by B. Halstead.		5-40
Synanceja verrucosa (Bloch & Schneider, 1801)	Recorded previously by B. Halstead.		0-20
DACTYLOPTERIDAE			
Dactyloptena orientalis (Cuvier, 1829)	Recorded previously by B. Halstead.		1-45
PLATYCEPHALIDAE			
Cymbacephalus beauforti Knapp, 1973	1, 11, 16, 17, 39,	A cryptic species that is rarely sighted.	2-12
Thysanophrys chiltoni Schultz, 1966	Seen on 1997 survey only.		1-80
SERRANIDAE			
Aethaloperca rogaa (Forsskål, 1775)	1, 5, 9, 10, 19, 24, 29, 31-35, 48	Occasional.	1-55
Anyperodon leucogrammicus (Valenciennes, 1828)	1, 4-6, 9, 11, 13, 14, 16-20, 22, 24, 27, 29-32, 34, 38, 39, 43, 45, 48-50, 53, 54	Moderately common, although always in low numbers.	5-50
Belonoperca chabanaudi Fowler and Bean, 1930	19, 44, 45	Occasional in caves on drop offs.	4-45
Cephalopholis argus Bloch and Schneider, 1801	1, 4, 5, 9, 12, 18, 21-23, 25, 27-30, 39, 43-45, 52	Occasional.	1-40
C. boenack (Bloch, 1790)	13, 15, 32, 34, 36, 40-42, 46	Occasional, in silty harbors and bays.	1-20
C. cyanostigma (Kuhl and Van Hasselt, 1828)	1-5, 9-34, 36, 40-42, 46, 54, 56, 57	Moderately common in more sheltered areas.	2-35
C. leopardus (Lacepède, 1802)	1, 2, 9, 14, 16, 19, 20, 23, 25-28, 30, 31, 37, 39, 43, 47, 49, 50, 53-55	Occasional.	3-25
C. microprion (Bleeker, 1852)	3-5, 7, 10-17, 20, 22-27, 29, 32, 33, 36, 40-42, 46, 56, 57	Occasional on relatively silty reefs.	2-20
C. miniata (Forsskål, 1775)	1, 5, 7, 9, 10, 13, 16, 18, 20-23, 28, 31-33, 35, 39, 42, 47-49, 51, 54	Moderately common, usually in areas of clear water.	3-150
C. polleni (Bleeker, 1868)	44	Rare, only one seen at 50 m depth.	20-120
C. sexmaculata Rüppell, 1828	1, 4, 9, 10, 19, 20, 24, 25, 28, 30, 37, 44	Occasional, on ceilings of caves on steep drop-offs.	6-140
C. sonnerati (Valenciennes, 1828)	Seen on 1997 survey only.	Occasional in deep water (below 20 m) of outer slopes.	10-100
C. spiloparaea (Valenciennes, 1828)	1, 2, 19, 27-30, 42, 44, 49, 50, 53	Occasional in deep water (below 20 m) of outer slopes.	16-108
C. urodeta (Schneider, 1801)	1, 4-6, 8-10, 16, 18-20, 29-31, 33, 35, 37, 39, 43-45, 47-50, 53-56	Moderately common in variety of habitats.	1-36
Cromileptes altivelis (Valenciennes, 1828)	13, 22, 30, 32, 33, 36, 39, 47, 48, 50	Occasional.	2-40
Diploprion bifasciatum Cuvier, 1828	5, 8, 32-34, 41, 55-57	Occasional, shletered inshore areas.	2-25

Species	Site records	Abundance	Depth (m)
Epinephelus areolatus (Forsskål, 1775)	Seen on 1997 survey only.		6-200
E. bontoides (Bleeker, 1855)	Recorded previously by B. Halstead.		2-30
E. caeruleopunctatus (Bloch, 1790)	17	Rare, only one adult seen.	5-25
E. coioides (Hamilton, 1822)	Seen on 1997 survey only.		2-100
E. corallicola (Kuhl and Van Hasselt, 1828)	15, 56, 57	Rare, only five individuals sighted. Photographed.	3-15
E. cyanopodus (Richardson, 1846)	32, 33, 36, 55, 57	Occasional in mainly sand-bottom areas.	5-150
E. fasciatus (Forsskål, 1775)	5, 9, 11, 14-16, 20, 22, 27, 29-32, 36, 49, 50, 54, 55	Moderately common, mainly in southern areas of MBP.	4-160
E. fuscoguttatus (Forsskål, 1775)	17, 20, 30, 35, 43, 48	Occasional.	3-60
E. hexagonatus (Bloch and Schneider, 1801)	Seen on 1997 survey only.		3-10
E. howlandi (Günther, 1873)*	45, 53	Rare, only two observed.	1-37
E. macrospilos (Bleeker)	56	Rare, only four individuals seen. Phtographed.	5-25
E. maculatus (Bloch, 1790)	33-36, 39, 47, 48,51	Occasional, mainly juveniles seen around debris on silty slopes.	10-80
E. merra Bloch, 1793	3, 4, 6-12, 14-19, 21, 26-29, 31, 32, 35, 37-54, 57	Moderately common, several seen on most dives.	1-15
E. ongus (Bloch, 1790)	2, 5, 7, 8, 15, 17, 23, 27, 33, 36	Occasional.	5-25
E. polyphekadion (Bleeker, 1849)	7, 13, 30, 37, 45, 51	Occasional.	2-45
E. quoyanus (Valenciennes, 1830)	33	Rare, only one individual seen.	1-10
E. spilotoceps Schultz, 1953	48	Rare, only one individual seen.	1-15
E. undulosus (Quoy and Gaimard, 1824)	31	Rare, only one individual seen.	10-90
Grammistes sexlineatus (Thünberg, 1792)	14	Rare, only one seen.	0.5-30
Grammistops ocellatus Schultz, 1953	22	Two specimens collected with rotenone.	5-30
Gracila albimarginata (Fowler and Bean, 1930)	2, 22, 23, 44, 52, 53	Occasional on outer slopes.	6-120
Liopropoma mitratum Lubbock & Randall, 1978*	25, 37	Two specimens collected with rotenone.	3-46
L. susumi (Jordan & Seale, 1906)*	53	One specimen collected with rotenone.	2-34
L. tonstrinum Randall and Taylor, 1988	22	One specimen collected with rotenone.	11-50
Luzonichthys waitei (Fowler, 1931)	Seen on 1997 survey only.		10-55
Plectranthias inermis Randall, 1980	Seen on 1997 survey only.		14-65
P. longimanus (Weber, 1913)	45, 53`	Two specimens collected with rotenone.	6-75
Plectropomus areolatus (Rüppell, 1830)	6, 17, 21	Rare, only three seen.	2-30
P. laevis (Lacepède, 1802)	4-9, 12, 18, 29-31, 34, 35, 37, 39, 43-45, 49, 50, 52-54	Moderately common. The most common coral trout in the Louisiades.	4-90

continued

Species	Site records	Abundance	Depth (m)
P. leopardus (Lacepède, 1802)	3, 5, 6, 8-11, 13, 15, 20, 22, 24, 29, 32-36, 38, 40-42, 46-48, 51, 55, 57	Moderately common. The most common coral trout in MBP.	3-100
P. maculatus (Bloch, 1790)	41	Rare, only one seen in silty area.	2-30
P. oligacanthus (Bleeker, 1854)	1, 2, 4-7, 9, 10, 12-14, 16, 20, 22, 24-29, 44, 57	Occasional.	4-40
Pseudanthias bicolor (Randall, 1979)	48	Rare, several individuals seen at one site.	5-68
P. cooperi (Regan, 1902)	Recorded previously by B. Halstead.		16-60
P. dispar (Herre, 1955)	1, 3-5, 19,48	Moderately common and locally abundant, but seen at few sites.	4-40
P. engelhardi Allen & Starck, 1982	Recorded previously by B. Halstead.		20-65
P. fasciatus (Kamohara, 1954)	Seen on 1997 survey only.		20-150
P. huchtii (Bleeker, 1857)	1, 2, 4-9, 13, 14, 16-35, 37-39, 43, 44, 48-50, 52-55, 57	Abundant, one of most common reef fishes in MBP. Photogrpahed.	4-20
P. hutomoi (Allen and Burhanuddin, 1976)	Seen on 1997 survey only.		30-60
P. hypselosoma Bleeker, 1878	4, 32	Rare, only a few seen at two sites.	10-40
P. luzonensis (Katayama and Masuda, 1983)	29, 30, 50, 52	Occasional, seen at only four sites in deep water (below 30 m), but locally common.	12-60
P. pleurotaenia (Bleeker, 1857)	2, 16, 19, 20, 26-29, 44, 49, 50, 52, 53, 55	Moderately common, on outer slopes below about 20 m depth. Photographed.	15-180
P. randalli (Lubbock & Allen, 1978)	Recorded previously by B. Halstead.		15-70
P. rubrizonatus (Randall, 1983)	Seen on 1997 survey only.		15-133
P. squamipinnis (Peters, 1855)	1, 4-7, 9, 18, 19, 22, 30, 39, 44, 45, 47-50, 52-55	Common, but usually less abundant than the similar P. huchtii.	4-20
P. tuka (Herre and Montalban, 1927)	1, 2, 4-6, 9, 10, 12-19, 21-31, 33-35, 37-39, 43-45, 49, 50, 52-54, 56, 57	Common in a variety of habitats, but usually in areas exposed to currents.	8-25
P. smithvanizi (Randall & Lubbock, 1981)	Recorded previously by B. Halstead.		6-70
P. sp	Recorded previously by B. Halstead.		50-70
Pseudogramma astigmum Randall and Baldwin, 1997	Seen on 1997 survey only.		10-46
P. polyacantha (Bleeker, 1856)	45	One specimen collected with rotenone.	1-15
Serranocirrhitus latus Watanabe, 1949	37, 44, 49, 52, 53		15-70
Variola albimarginata Baissac, 1953	1, 6, 7, 16, 19, 22, 26, 29-31, 33, 35, 49, 50, 52-55	Occasional and always in low numbers.	12-90
V. louti (Forsskål, 1775)	5, 22, 29, 30, 39, 44, 45, 47, 48, 50, 52, 54	Occasional and always in low numbers.	4-150

continued

Species	Site records	Abundance	Depth (m)
PSEUDOCHROMIDAE			
Cypho purpurescens (De Vis, 1884)	45	One collected with rotenone.	5-35
Pseudochromis bitaeniatus (Fowler, 1931)	4, 6, 12-14, 22, 24, 25, 28, 33, 39	Occasional, among crevices and ledges.	5-30
P. fuscus (Müller and Troschel, 1849)	3, 4, 7, 8, 10-12, 14, 15, 18, 21-24, 32, 34-36, 41, 46, 51, 56	Occasional, around small coral and rock outcrops.	1-30
P. marshallensis (Schultz, 1953)	12, 13, 20, 32, 35, 41, 45	Occasional under rocky overhangs.	2-25
P. paccagnellae Axelrod, 1973	1, 6, 12, 13, 15, 18-20, 22, 24-26, 29, 30, 35, 37-39, 43-45, 52, 53, 57	Occasional.	6-70
P. paranox Lubbock and Goldman, 1976	5, 9, 13, 20, 22-24, 34, 50, 56, 57	Occasional, under ledges and among rubble.	5-20
P. perspicillatus Günther, 1862	4	Only one seen in 20 m.	3-20
P. sp. 1	Seen on 1997 survey only.		5-25
P. sp. 2	3, 9	Rare, only two seen below 30 m.	20-50
P. tapienosoma Bleeker, 1853	15	Only one seen, but has cryptic habits.	2-60
Pseudoplesiops annae (Weber, 1913)	37	Several collected with rotenone.	4-25
P. knighti Allen, 1987	37	Collected with rotenone.	5-35
P. multisquamatus Allen, 1987	22, 24, 25	Collected with rotenone.	5-35
P. rosae Schultz, 1943	Recorded previously by B. Halstead.		8-40
P. typus Bleeker, 1858	Recorded previously by B. Halstead.		5-30
PLESIOPIDAE			
Assessor flavissimus Allen and Kuiter, 1976	1, 6, 8, 10, 12-14, 24-26, 37, 38, 43	Moderately common, seen at several sites under ledges.	5-20
Calloplesiops altivelis (Steindachner, 1903)	45	One collected with rotenone.	3-45
Plesiops corallicola Bleeker, 1853	Seen on 1997 survey only.		0-3
Steeneichthys plesiopsus Allen and Randall, 1985	Seen on 1997 survey only.		3-40
CIRRHITIDAE			
Cirrhitichthys aprinus (Cuvier, 1829)	Seen on 1997 survey only.		5-40
C. falco Randall, 1963	9, 20, 24, 26, 30, 43, 45, 47-50	Occasional. Photographed.	4-45
C. oxycephalus (Bleeker, 1855)	5, 6, 16, 19, 26, 47, 55	Occasional.	2-40
Cyprinocirrhites polyactis (Bleeker, 1875)	6	Rare.	10-132
Neocirrhites armatus Castelnau, 1873	Recorded previously by B. Halstead.		1-15
Oxycirrhitus typus Bleeker, 1857	Seen on 1997 survey only.		10-100
Paracirrhites arcatus (Cuvier, 1829)	4, 5, 9, 14, 17, 19, 25, 27, 29, 39, 43-45, 47-50, 52-55	Moderately common, but in lower numbers than P. forsteri. Photographed.	1-35

continued

Species	Site records	Abundance	Depth (m)
P. forsteri (Schneider, 1801)	1, 2, 4-7, 16-20, 22-30, 33, 37-39, 43-45, 47-50, 52-55, 57	Moderately common, the most abundant hawkfish in MBP; seen on regular basis, but in relatively low numbers. Photographed.	1-35
OPISTOGNATHIDAE			
Opistognathus sp. 1	38	Rare, only two fish seen.	5-20
O. sp. 2	Recorded previously by B. Halstead.		5-20
TERAPONTIDAE			
Mesopristes argenteus (Cuvier, 1829)	Recorded by GRA in 1999.		0-4
Terapon jarbua (Forsskål, 1775)	Seen on 1997 survey only.		0-5
PRIACANTHIDAE			
Priacanthus blochii Bleeker, 1853	Seen on 1997 survey only.		15-30
P. hamrur (Forsskål, 1775)	36, 56	Rare, only five seen.	5-80
APOGONIDAE			
Apogon abogramma Fraser and Lachner, 1985	24	Rare, about 10 individuals seen in caves below 30 m depth.	20-40
A. angustatus (Smith and Radcliffe, 1911)	Seen on 1997 survey only.		5-30
A. apogonides (Bleeker, 1856)	Seen on 1997 survey only.		12-40
A. aureus (Lacepède, 1802)	30, 31, 51	Rare, only a few scattered fish seen at three sites.	10-30
A. bandanensis Bleeker, 1854	14, 15, 27, 38, 42, 57	Occasional.	3-10
A. cirysotaenia Bleeker, 1851	Seen on 1997 survey only.		1-14
A. compressus (Smith and Radcliffe, 1911)	4, 7, 8, 11, 14, 15, 17, 20, 21, 24-28, 33, 35, 36, 38, 41, 45, 46, 49, 51, 54, 56, 57	Moderately common, one of most abundant cardinalsfishes seen during the day, usually among branching Acropora corals.	2-20
A. crassiceps Garman, 1903	53	Collected with rotenone. Reported previously as A. coccineus.	1-30
A. cyanosoma Bleeker, 1853*	29, 39, 45, 48	Occasional. Phtographed.	3-15
A. dispar Fraser and Randall, 1976	Seen on 1997 survey only.		12-50
A. evermanni Jordan and Snyder	24	Rare, one unusually large fish (about 150 mm TL) seen on roof of cave in 20 m.	10-40
A. exostigma Jordan and Starks, 1906	29, 42, 51	Rare, several fish seen at three sites.	3-25
A. fleurieu (Lacepède, 1802)	Seen on 1997 survey only.		5-30
A. fraenatus Valenciennes, 1832	4, 14, 21, 26, 27, 32, 33, 38, 39, 42, 44, 46, 50, 51	Seen at relatively few sites, but locally common under ledges and in coral crevices.	3-35

continued

Species	Site records	Abundance	Depth (m)
A. fragilis Smith, 1961	10, 11, 21, 40, 42, 51	Rarely seen, but locally abundant.	1-15
A. fuscus Quoy and Gaimard, 1824	11	Rarely seen during day, but probably common throughout MBP.	3-15
A. guamensis Valenciennes, 1832	Seen on 1997 survey only.		0-8
A. hartzfeldi Bleeker, 1852	Seen on 1997 survey only.		1-10
A. hoeveni Bleeker, 1854	3, 36	Rarely seen, but locally common; sometimes sheltering in Diadema sea urchins.	1-25
A. kallopterus Bleeker, 1856	5, 14, 33, 35, 41, 42, 44, 51	Rarely seen, due to nocturnal habitats.	3-35
A. kiensis Jordan & Snyder, 1801	Recorded previously by B. Halstead.		5-40
A. leptacanthus Bleeker, 1856	21	Rarely encountered, but locally common among branching corals.	1-12
A. melanoproctus Fraser and Randall, 1976	22, 24, 25, 57	Rarely seen, but locally common in caves on steep slopes. The species previously reported from MBP as A. dispar is possibly A. melnaproctus. Collected with rotenone at site 22.	15-40
A. melas Bleeker, 1848	Recorded previously by B. Halstead.		3-15
A. moluccensis Valenciennes, 1832	51	Rare, a single fish seen.	3-35
A. multilineatus Bleeker, 1865	14, 15, 32, 57	Rare, but nocturnal habits.	1-5
A. nanus Allen, Kuiter, and Randall, 1994	Seen on 1997 survey only.		5-20
A. neotes Allen, Kuiter, and Randall, 1994	4, 10, 11, 14, 15, 20, 22, 26, 32, 34, 40, 41, 42, 46, 56, 57	Occasional, but locally common, often adjacent to steep slopes around black coral.	10-25
A. nigrofasciatus Schultz, 1953	1, 5, 6, 9, 12-14, 16-19, 21, 23-28, 31, 32, 35, 37-39, 42-45, 47-55, 57	Moderately common, one of most abundant cardinalfishes, but always in small numbers under ledges and among crevices.	2-35
A. notatus (Houttuyn, 1782)	Seen on 1997 survey only.		2-30
A. novemfasciatus Cuvier, 1828	7	Several pairs seen in shallows.	0.5-3
A. ocellicaudus Allen, Kuiter, and Randall, 1994	11, 41, 57	Generally rare, but locally common at three sites. Photographed.	11-55
A. perlitus Fraser & Lachner, 1985	Recorded previously by B. Halstead.		2-15
A. properupta (Whitley, 1964)	4, 5, 31, 33, 34, 51, 56	Rare, a few seen at only one site. Photographed.	1-15
A. quadrifasciatus Cuvier, 1828	23	Rare, one seen on sand slope in 20 m.	5-40
A. rhodopterus Bleeker, 1852	Recorded previously by B. Halstead.		10-40
A. sealei Fowler, 1918*	10, 33, 39, 40, 42, 51	Occasional.	2-12
A. selas Randall and Hayashi, 1990	22, 27, 38, 40, 41	Rarely encountered, but locally common.	20-35

continued

Species	Site records	Abundance	Depth (m)
A. similis Fraser, Randall, & Lachner, In press*	46	A few seen at one site.	2-12
A. semiornatus Peters, 1876	Recorded previously by B. Halstead.		5-35
A. taeniophorus Regan, 1908	11, 15, 47	Relatively rare, but occurs in very shallow water and is nocturnal and therefore difficult to accurately survey.	0.5-2
A. talboti Smith, 1961	45, 53	Several collected with rotenone. Formerly reported as A. unicolor.	10-30
A. thermalis Cuvier, 1829	21	Common at one site. Formerly reported as A. sangiensis.	0-10
A. trimaculatus Cuvier, 1828	28, 40, 41, 57	Rare, but difficult to survey due to nocturnal habitats.	2-10
A. ventrifasciatus Allen, Kuiter, and Randall, 1994	Seen on 1997 survey only.		5-35
Archamia biguttata Lachner, 1951	24	A single school seen in a large cave.	5-18
A. fucata (Cantor, 1850)	9, 11, 13, 21, 24, 29, 32, 34, 38, 40, 42, 46, 51	Occasional, but common at several sites.	3-60
A. macropterus (Cuvier, 1828)	Seen on 1997 survey only.		3-15
A. zosterophora (Bleeker, 1858)	8, 10, 11, 13, 21, 38, 40, 42, 56, 57	Occasional, but common at some sites.	2-15
Cercamia eremia (Allen, 1987)	53	Collected with rotenone.	5-40
Cheilodipterus alleni Gon, 1993	8, 12, 14, 18, 20, 24, 26, 57	Occasional, especially in caves and crevices on steep slopes.	1-25
C. artus Smith, 1961	4, 7, 8, 10-15, 21, 22, 24, 27, 28, 33, 34, 38, 40, 42, 50, 51, 56, 57	Moderately common, often among branching corals.	2-20
C. isostigmus (Schultz, 1940)	13, 14	Difficult to distinguish from C. quinquelineatus, but two specimens speared for identification.	1-20
C. intermedius Gon, 1993*	30, 39, 40	Found at only three sites, but locally common. Photographed.	4-15
C. macrodon Lacepède, 1801	11, 13, 14, 24, 25, 27, 29, 31, 32, 34, 35, 46, 50, 51, 53, 55-57	Moderately common, but always in low numbers (except juveniles).	4-30
C. parazonatus Gon, 1993	4, 7, 10-15, 20, 34, 36, 40, 42, 46, 51, 56, 57	Moderately common, but seen in small numbers on sheltered inshore reefs.	1-35
C. quinquelineatus Cuvier, 1828	3, 4, 7, 10-15, 17-39, 41-46, 48-52, 54, 56, 57	Common, most abundant member of genus in MBP.	1-40
C. singapurensis Bleeker, 1859	Recorded previously by B. Halstead.		2-15
Foa brachygramma (Jenkins, 1902)*	42	One seen, but very cryptic and difficult to assess.	1-15
Fouleria aurita (Valenciennes, 1831)	22, 25, 37	Several specimens collected with rotenone.	0-15
F. vaiulae (Jordan and Seale, 1906)	Seen on 1997 survey only.	Formerly reported as Fouleria abocellata.	3-20
Gymnapogon sp.	Recorded previously by B. Halstead.		1-15
G. urospilotus Lachner, 1953	Recorded previously by B. Halstead.		1-15

continued

Species	Site records	Abundance	Depth (m)
Pseudamia gelatinosa Smith, 1955	Recorded previously by B. Halstead.		1-40
P. hayashi Randall, Lachner and Fraser, 1985	Seen on 1997 survey only.		2-64
P. zonata Randall, Lachner & Fraser, 1985	Recorded previously by B. Halstead.		10-35
Rhabdamia cypselurus Weber, 1909	11, 18, 31, 32, 34, 40	Occasionally observed, but sometimes in high numbers swarming around coral bommies.	2-15
R. gracilis (Bleeker, 1856)	34, 42	Rarely observed, but sometimes in high numbers swarming around coral bommies.	5-20
R. sp.	Seen on 1997 survey only.		10-25
Siphamia corallicola Allen, 1993*	36	Common at one site in 10-20 m. Photographed.	5-30
S. majimae Matsubara & Iwai, 1958*	36	A group of about 20 fish seen with sea urchin. Photographed.	5-40
S. versicolor (Smith and Radcliffe, 1912)	Seen on 1997 survey only.		3-25
Sphaeramia nematoptera (Bleeker, 1856)	15, 21, 56, 57	Occasional, but locally common among sheltered corals. Photographed.	1-8
S. orbicularis (Cuvier, 1828)	Seen on 1997 survey only.		0-3
SILLAGINIDAE			
Sillago sihama (Forsskål, 1775)	Recorded previously by B. Halstead.		0-15
MALACANTHIDAE			
Hoplolatilus cuniculus Randall & Dooley, 1974	6, 30	Common below 50 m on rubble slopes.	25-115
H. marcosi Burgess, 1978	Recorded previously by B. Halstead.		18-80
H. pohlei Earle & Pyle, 1997	Recorded previously by B. Halstead.		25-70
H. purpureus Burgess, 1978	Recorded previously by B. Halstead.		18-80
H. starcki Randall and Dooley, 1974	2, 19, 30, 52, 53	Rarely encountered as their distribution lies mainly below the depths of sampled during the survey.	20-105
Malacanthus brevirostris Guichenot, 1848	30, 39, 45, 47, 54	Occasional.	10-45
M. latovittatus (Lacepède, 1798)	1, 5, 12, 19, 29, 44, 45, 47, 48, 54	Occasional.	5-30
ECHENEIDAE			
Echeneis naucrates Linnaeus, 1758	43, 44	A few individuals seen attached to sharks.	0-30
CARANGIDAE			
Alepes vari (Cuvier, 1833)	Seen on 1997 survey only.		2-50
Atule mate (Cuvier, 1833)	9, 31, 34, 41, 57	Several schools containing approximately 20-50 fish seen.	2-50
Carangoides bajad (Forsskål, 1775)	1-3, 6, 9, 10, 12, 13, 16-25, 27-30, 33, 44, 56	Occasional, usually in low numbers.	5-30

continued

Species	Site records	Abundance	Depth (m)
C. ferdau (Forsskål, 1775)	22, 33, 36, 46, 47, 51	Occasional in small schools, apparently feeding on benthic invertebrates on sandy bottoms.	2-40
C. fulvoguttatus (Forsskål, 1775)*	4-6, 14, 31, 32, 40-42, 51, 54	Occasional. Photographed.	5-100
C. oblongus (Cuvier, 1833)	Seen on 1997 survey only.		5-40
C. orthogrammus (Jordan and Gilbert, 1882)	Seen on 1997 survey only.		3-168
C. plagiotaenia Bleeker, 1857	4-6, 8, 9, 13-15, 17, 20, 22, 28, 30, 31, 33, 34, 38, 39, 41-46, 49, 52, 53, 56, 57	Moderately common, but usually solitary fish encountered.	5-200
Caranx ignobilis (Forsskål, 1775)	18, 27	Rare, two large adults seen.	2-80
C. melampygus Cuvier, 1833	4-6, 9, 11, 13, 18, 19, 21, 28-30, 34, 36, 37, 39, 40, 42-50, 52-54	Moderately sommon, ususaly seen solitary or in small schools.	1-190
C. papuensis Alleyne & Macleay, 1877	Recorded previously by B. Halstead.		1-50.
C. sexfasciatus Quoy and Gaimard, 1825	2, 4, 44	Rarely seen, and ususally in small to large schools (approximately 5-500 fish).	3-96
Elegatis bipinnulatus (Quoy and Gaimard, 1825)	1, 2, 6, 12, 37, 38, 41, 44, 47	Moderately common adjacent to steep outer slopes. Photographed.	5-150
Gnathanodon speciosus (Forsskål, 1775)	33, 35, 46, 51	Rare, but often found some distance from reefs on flat sand bottoms.	1-30
Scomberoides lysan (Forsskål, 1775)	18, 19, 47	Rare, only one seen.	1-100
S. tol (Cuvier, 1832)	2	Rare, only one seen.	1-100
Selar boops (Cuvier, 1833)	Seen on 1997 survey only.		1-30
S. crumenophthalmus (Bloch, 1793)	51	A small school seen next to Panasia Island.	1-170
Selaroides leptolepis (Kuhl and van Hasselt, 1833)	Seen on 1997 survey only.		1-25
Trachinotus baillonii (Lacepède, 1801)	7	Rare, one school with about 20 fish seen in sandy shallows.	0.5-5
T. blochii (Lacepède, 1801)	4	Rare, one school with about 10 fish seen.	3-40
LUTJANIDAE			
Aphareus furca (Lacepède, 1802)	37, 39, 43-45, 52, 54	Occasional in Louisiades.	6-70
Aprion virescens Valenciennes, 1830	9, 30, 31, 35, 39, 44, 45	Occasional, mainly in Louisiades.	3-40
Lutjanus argentimaculatus (Forsskål, 1775)	Seen on 1997 survey only.		1-100
L. biguttatus (Valenciennes, 1830)	1, 7, 8, 11-15, 21, 22, 28, 29, 32-34, 36-38, 40, 42, 46, 49, 55-57	Occasional, mainly on sheltered reefs with rich corals.	3-40
L. bohar (Forsskål, 1775)	1, 2, 4-6, 9, 10, 14, 16-22, 24-26, 28-35, 37-39, 41, 43-56	Common, one of the three most common snappers, but usually in relatively low numbers at each site.	4-180

continued

Species	Site records	Abundance	Depth (m)
L. boutton (Lacepède, 1802)	3, 27	Rarely seen and usually in low numbers.	5-25
L. carponotatus (Richardson, 1842)	3, 7, 11-13, 15, 34, 36, 38, 41, 42, 46, 57	Occasional. Usually on sheltered coastal reefs.	2-35
L. ehrenburgi (Peters, 1869)	21, 24, 51	Rarely seen, but locally common near mangrove shores.	1-20
L. fulviflamma (Forsskål, 1775)	16, 28, 55, 56	Rarely seen, but locally common at a few sites.	1-35
L. fulvus (Schneider, 1801)	1, 7, 13, 17, 22, 24, 27-29, 32, 38, 42, 43, 46, 49, 51, 52, 54, 56, 57	Moderately commom, but usually in small numbers.	2-40
L. gibbus (Forsskål, 1775)	3, 4, 6, 8-14, 16-27, 29-31, 33-38, 43-45, 47-54, 57	Common, one of three most common snappers. An extraordinarily large school containing hundreds of fish seen at Bramble Haven (site 54).	6-40
L. kasmira (Forsskål, 1775)	29, 39, 43, 48, 50, 55	Occasional, ususally in low numbers, but schools present at site 55. Photographed.	3-265
L. lutjanus Bloch, 1790*	11	Rare.	10-90
L. madras (Valenciennes, 1831)	Seen on 1997 survey only.		5-90
L. mizenkoi Allen & Talbot, 1985*	15, 27	Rare, several individuals of this rare species seen at two sites. Photographed.	15-80
L. monostigma (Cuvier, 1828)	1, 4, 6, 19, 20, 22, 25, 26, 30-34, 37, 39, 42-45, 52, 53	Moderately common, between 10- 20 seen on some dives.	5-60
L. quinquelineatus (Bloch, 1790)	4, 5, 7, 8, 11, 21, 22, 31-34, 42, 46, 51, 55, 57	Occasional, usually in small numbers, but large aggregation seen at Bently Island.	5-30
L. rivulatus (Cuvier, 1828)	4, 14, 19-23, 27, 44, 47	Occasional, a few large adults and several half-grown fish seen.	2-100
L. rufolineatus (Valenciennes, 1830)	21	Rare.	12-50
L. russelli (Bleeker, 1849)	14, 17-20, 22, 43, 57	Occasional solitary fish sighted.	1-80
L. sebae (Cuvier, 1828)	Recorded previously by B. Halstead.		10-100
L. semicinctus Quoy and Gaimard, 1824	1-6, 8, 11-31, 25-39, 42, 43, 46, 47, 49, 52-54, 56, 57	Common, one of the three most common snappers, but usually in relatively low numbers at each site.	10-40
L. timorensis (Quoy and Gaimard, 1824)	Seen on 1997 survey only.		6-130
L. vitta (Quoy and Gaimard, 1824)	33, 36, 42, 56	Occasional at sheltered coastal sites.	8-40
Macolor macularis Fowler, 1931	1-3, 5-9, 12, 14, 16-20, 22-39, 41, 43-45, 47, 48, 50, 52-55	Common.	3-50
M. niger (Forsskål, 1775)	4, 6, 14, 15, 21, 22, 29-31, 34, 35, 39, 40, 43, 45, 48-50, 54	Moderately common, but in much lower numbers than M. macularis.	3-90
Paracaesio sordidus Abe & Shinohara, 1962*	2, 19, 24	Several schools sighted on steep slopes, usually below 30 m depth.	5-100

continued

continued

Species	Site records	Abundance	Depth (m)
P. xanthurus (Bleeker, 1869)	Seen on 1997 survey only.	Rare, a school of about 30 fish seen.	20-50
Symphorichthys spilurus (Günther, 1874)	5, 21, 31, 35, 46, 51, 56, 57	Rare, a few adults seen. Photographed.	5-60
Symphorus nematophorus (Bleeker, 1860)	9, 31, 36	Rare, a few adults seen.	5-50
CAESIONIDAE			
Caesio caerulaurea Lacepède, 1802	2, 4-10, 12-14, 16, 19, 21, 23-26, 28-30, 32-38, 40-42, 44-46, 48, 52-57	Abundant in variety of habitats.	1-30
C. cuning (Bloch, 1791)	1, 2, 5-13, 15-17, 19-23, 25-28, 32-38, 41, 42, 44, 46, 52, 56, 57	Abundant in variety of habitats, particularly coastal reefs.	1-30
C. lunaris Cuvier, 1830	3-6, 13, 14, 18, 21, 24, 26, 29, 30, 44. 47	Occasional, but locally common.	1-35
C. teres Seale, 1906	1, 2, 4, 5, 23, 33, 38, 44, 54	Occasional, but locally common.	1-40
Gymnocaesio gymnoptera (Bleeker, 1856)	1, 5, 6, 13, 44, 46	Occasionally seen with mixed school of fusiliers, mainly Pterocaesio pisang.	5-30
Pterocaesio digramma (Bleeker, 1865)	2, 3, 7, 9, 10, 12, 13, 16, 20, 22, 28, 55-57	Common in northern parts of MBP. Photographed.	1-25
P. lativittata Carpenter, 1987*	44	One school of about 200 fish seen in 50 m depth.	10-70 m
P. marri Schultz, 1953	30, 32, 35,38, 41, 43, 44, 48-50, 52	Common in Louisiades.	1-35
P. pisang (Bleeker, 1853)	1, 3-14, 17-24, 28-30, 32, 33, 35, 37-39, 42, 44-46, 49-54, 56, 57	Common in variety of habitats.	1-35
P. tessellata Carpenter, 1987	29, 30, 37, 44, 53	Occasional, but locally abundant.	1-35
P. tile (Cuvier, 1830)	1, 4-6, 21, 29-31, 43, 45, 48, 52-56	Occasional, but locally abundant.	1-60
P. trilineata Carpenter, 1987	3, 4, 6, 7, 9, 14, 29, 30, 32, 35, 37-39, 42, 44, 45, 48-51, 53, 54, 57	Moderately common, but locally abundant.	1-30
GERREIDAE			
Gerres abbreviatus Bleeker, 1850	Seen on 1997 survey only.		0-40
G. argyreus (Schneider, 1801)	Recorded previously by B. Halstead.		0-10
G. filamentosus Cuvier, 1829	Recorded previously by B. Halstead.		0-10
G. oyena (Forsskål, 1775)	Seen on 1997 survey only.		0-10
HAEMULIDAE			
Diagramma pictum (Thünberg, 1792)	33, 56	Rare, a few seen in silty areas.	2-40
Plectorhinchus celebicus Bleeker, 1873	5, 14, 23, 34, 38, 39, 42, 46, 51, 56, 57	Occasional.	6-30
P. chaetodontoides (Lacepède, 1800)	2, 5-11, 13-15, 17, 18, 20, 22, 28-30, 32, 34-39, 41, 43, 44, 46, 50, 51, 54, 57	Moderately common, the most abundant sweetlips in MBP, but always seen in small numbers.	1-40
P. gibbosus (Lacepède, 1802)	30, 32, 44, 45	Rare, only four adults seen.	2-30

Species	Site records	Abundance	Depth (m)
P. lessoni (Cuvier, 1830)	6, 29, 39, 54, 55	Rare, only five seen.	5-35
P. lineatus (Linnaeus, 1758)	4, 6, 7, 12, 19, 23, 30, 31, 36, 39, 40, 44, 45, 47, 49	Occasional.	2-40
P. obscurus (Günther, 1871)	20, 43-45	Rare, four large adults seen.	5-50
P. orientalis (Bloch, 1793)	19, 26, 27, 31, 32, 40, 55	Occasional.	3-30
P. pictus (Cuvier, 1830)	28, 30	Rare, only two seen.	3-50
LETHRINIDAE			
Gnathodentex aurolineatus Lacepède, 1802	26, 30, 45, 51, 53, 54	Rarely encountered, but sometimes locally common.	1-30
Gymnocranius grandoculus (Valenciennes, 1830)	31, 32, 35	Rare, except group of 100 seen at site 31 on sand bottom in 15-20m. Photographed.	20-100
G. sp.	43, 45, 51, 54	Occasional.	15-40
Lethrinus atkinsoni Seale, 1909	3, 15, 31, 55	Rarely seen, usually in mixed reef-sand areas.	2-30
L. erythracanthus Valenciennes, 1830	1, 4, 7, 8, 11, 12, 14, 29, 35, 37, 43, 44, 50	Both juveniles and adults occasionally sighted.	15-120
L. erythropterus Valenciennes, 1830	4, 12, 14, 16, 17, 19-22, 25, 26, 29, 33, 35, 36, 40, 41, 43, 45, 48, 49, 57	Both juveniles and adults occasionally sighted.	2-30
L. genivittatus Valenciennes, 1830	Recorded previously by B. Halstead.		5-25
L. harak (Forsskål, 1775)	3, 7, 11, 12, 14, 21, 32, 36, 38, 42, 51, 56	Moderately common, second most abundant member of family in MBP, commonly seen in shallow waters with sand or rubble bottoms.	1-20
L. lentjan (Lacepède, 1802)	Seen on 1997 survey only.		10-50
L. microdon Valenciennes, 1830	Seen on 1997 survey only.		10-50
L. nebulosus (Forsskål, 1775)	14, 31, 51, 56	Rarely seen, but common on sand bottom at site 31.	5-75
L. obsoletus (Forsskål, 1775)	3, 8, 10, 11, 18, 21, 29, 32, 45, 50, 54-56.	Occasional, and always in low numbers.	1-25
L. olivaceous Valenciennes, 1830	5, 6, 9, 19, 21, 25, 30, 35, 39, 44-46, 51	Occasional, in low numbers.	4-185
L. ornatus Valenciennes, 1830	3, 56	Rare.	3-20
L. semicinctus Valenciennes, 1830	Seen on 1997 survey only.		10-40
L. variegattus Valenciennes, 1830	3	Rare, but seagrass is main habitat.	1-10
L. xanthocheilus Klunzinger, 1870	29, 30, 45, 47	Rare, less than 10 fish seen.	2-25
Monotaxis grandoculis (Forsskål, 1775)	1-57	Common. The most abundant lethrinid in MBP.	1-100
NEMIPTERIDAE			
Pentapodus caninus (Cuvier, 1830)*	6, 11, 13, 56	Occasional in sand-rubble areas. Photographed.	15-60
Pentapodus emeryii (Richardson, 1843)	Seen on 1997 survey only.		5-40
P. paradiseus (Günther, 1859)*	55	Rare, only one seen.	3-30

continued

Species	Site records	Abundance	Depth (m)
P. sp.	1, 7-10, 12, 15, 17, 20, 22, 26, 27, 29, 32, 33, 35, 38, 41, 55, 57	Occasional, but locally common at several sites.	3-25
P. trivittatus (Bloch, 1791)	10, 15, 21, 33, 40, 42, 46, 56	Occasional, usually on sheltered coastal reefs.	1-15
Scolopsis affinis Peters, 1876	26, 32, 33, 51, 56	Occasional, but locally common in sandy areas.	3-60
S. bilineatus (Bloch, 1793)	1-10, 12-14, 16-40, 42-57	Common.	2-20
S. ciliatus (Lacepède, 1802)	3, 14, 15, 17, 18, 21, 36, 40-42, 56, 57	Moderately common at sites subjected to silting.	1-30
S. lineatus Quoy and Gaimard, 1824	10, 11, 14, 21, 27, 38, 53	Occasional on shallow reefs.	0-10
S. margaritifer (Cuvier, 1830)	3-5, 7, 8, 10-17, 20-23, 26, 28, 31-42, 46, 51, 56, 57	Common, especially on sheltered coastal reefs	2-20
S. monogramma (Kuhl and Van Hasselt, 1830)	Seen on 1997 survey only.		5-50
S. temporalis (Cuvier, 1830)*	15, 21, 23, 32, 33, 35, 36, 42, 47, 51, 56	Occasional, locally common over sand bottoms.	5-30
S. trilineatus Kner, 1868	4, 39	Rare.	1-10
S. xenochrous (Günther, 1872)	1, 6, 16	Generally rare, but not uncommon at 3 sites.	5-50
MULLIDAE			
Mulloidichthys flavolineatus (Lacepède, 1802)	10, 21, 23, 26, 28, 38, 42, 48, 51-53, 56	Occasional, but sometimes locally common.	1-40
M. vanicolensis (Valenciennes, 1831)	12, 32, 42, 51	Rarely seen and usually in low numbers.	1-113
Parupeneus barberinoides (Lacepède, 1801)	3, 14	Rare.	1-20
P. barberinus (Lacepède, 1801)	1-5, 7, 8, 10, 12-23, 26, 28-57	Common, the most abundant goatfish in MBP. Photographed.	1-100
P. bifasciatus (Lacepède, 1801)	1-22, 25-35, 37-57	Common, particularly on outer slopes and adjacent reefs.	1-80
P. cyclostomus (Lacepède, 1802)	1-8, 10, 15, 20, 21, 26, 29-34, 37-39, 41-45, 47-50, 52-55, 57	Moderately common, but in lower numbers than previous two species.	2-92
P. heptacanthus (Lacepède, 1801)	21, 22, 26, 27	Rare.	1-60
P. indicus (Shaw, 1903)	3, 36, 56	Rare, usually seen in silted areas.	0-15
P. macronema (Lacepede, 1802)	Recorded previously by B. Halstead.		5-30
P. multifasciatus Bleeker, 1873	1-22, 26, 29-35, 37-57	Common, second most abundant goatfish in MBP, consistently seen at most sites.	1-140
P. pleurostigma (Bennett, 1830)	29, 30, 48, 49, 54	Rare, confined to clean, white sand near reefs.	5-46
Upeneus tragula Richardson, 1846*	23, 38	Rare, but mainly found on sand bottoms away from reefs.	1-40
PEMPHERIDAE			
Parapriacanthus ransonneti Steindachner, 1870	29, 31, 32	Rarely encountered, but forms dense aggregations.	5-30
Pempheris adusta Bleeker, 1877	Seen on 1997 survey only.		5-30

continued

Species	Site records	Abundance	Depth (m)
P. mangula Cuvier, 1829*	24, 26	Two groups seen in caves. Has black outer margin on dorsal and anal fins.	5-30
P. oualensis Cuvier, 1831	22, 30, 36, 52, 53, 56	Probably common, but difficult to survey due to cryptic diurnal behaviour. Has black spot covering pectoral-fin base.	3-38
TOXOTIDAE			
Toxotes jaculatrix (Pallas, 1767)	Seen on 1997 survey only.		0-2
KYPHOSIDAE			
Kyphosus cinerascens (Forsskål, 1775)	4, 11, 31, 44, 52, 56	Occasional, but sometimes locally common.	1-24
K. vaigiensis (Quoy and Gaimard, 1825)	4, 8, 10, 12, 14, 22, 28, 31, 41, 42, 51, 53	Occasional, but sometimes locally common.	1-20
MONODACTYLIDAE			
Monodactylus argenteus (Linnaeus, 1758)	Seen on 1997 survey only.		0-5
CHAETODONTIDAE			
Chaetodon auriga Forsskål, 1775	4, 6, 10, 11, 14, 20, 21, 22, 29, 30-32, 34-40, 42, 45-49, 51, 54-57	Occasional, ususally areas with weed and sand mixed with coral reef.	1-30
C. baronessa Cuvier, 1831	1-6, 8-40, 43, 44, 46-57	Common, seen on nearly every dive.	2-15
C. bennetti Cuvier, 1831	1-6, 14, 19, 23, 25-28, 32, 35-37, 46-48, 52, 53, 56, 57	Occasional, frequently on outer slopes. Photographed.	5-30
C. burgessi Allen & Starck, 1973	Recorded previously by B. Halstead.		20-100
C. cirrinellus Cuvier, 1831	3-8, 10, 12, 14, 18, 20, 29-33, 36-40, 43-45, 47-55	Common on shallow reefs affected by surge.	1-12
C. ephippium Cuvier, 1831	1, 5, 6, 8, 10, 12-15, 17-22, 29-31, 33, 35-41, 43-57	Occasional, never more than 2-3 pairs seen at a single site.	1-30
C. kleinii Bloch, 1790	1-7, 9, 10, 19, 26, 29-32, 35, 37, 39, 40, 42-56	Cmmonly seen at most sites.	6-60
C. lineolatus Cuvier, 1831	8, 10, 20, 33, 34, 36, 46, 56, 57	Occasional, less common than the very similar C. oxycephalus.	2-170
C. lunula Lacepède, 1803	3, 4, 11, 14, 18, 21, 22, 25, 26, 36, 39, 45, 47, 52	Occasional.	1-40
C. lunulatus Quoy and Gaimard, 1824	1-4, 6-10, 12-30, 33-57	Common, one of the most abundant butterflyfishes in MBP; seen on almost every dive.	1-25
C. melannotus Schneider, 1801	4, 6, 9, 12, 28, 29, 31, 37, 39, 40, 43-45, 47, 48, 51-53	Occasional.	2-15
C. mertensii Cuvier, 1831*	1, 2, 14, 16-20, 22, 24, 26-28, 30, 36, 38, 43, 46, 53, 57	Occasional, more common at mainland and northern sites in MBP.	10-120
C. meyeri Schneider, 1801	43	Rare, only one seen.	5-25
C. ocellicaudus Cuvier, 1831	8, 11, 12, 14, 15-17, 22, 24, 26-28, 35, 36, 51, 54, 56, 57	Moderately common, seen at the majority of sites.	1-15
C. octofasciatus Bloch, 1787	3, 8, 10-15, 33-36, 40-42, 46, 56, 57	Occasional, except common at a fewinshore sites where reef influenced by silt.	3-20

continued

List of the reef fishes of Milne
Bay Province, Papua New Guinea

continued

Species	Site records	Abundance	Depth (m)
C. ornatissimus Cuvier, 1831	2, 4-9, 11, 14, 16-19, 21, 25, 27, 29, 30, 37-39, 43-45, 47-50, 53-55	Moderaely common in rich coral areas.	1-36
C. oxycephalus Bleeker, 1853	8, 11, 14, 15, 17, 19, 22, 25, 28-30, 41, 56, 57	Moderately common, but always in low numbers.	8-30
C. pelewensis Kner, 1868	5, 29-31, 37-39, 43-45, 48-50, 52-54	Common in the Louisiades. Photographed.	6-45
C. plebeius Cuvier, 1831	2, 5, 10, 11, 17-21, 29, 30, 31, 33, 35, 37, 39, 43-52, 54, 55	Occasional.	1-15
C. punctatofasciatus Cuvier, 1831	1-7, 9, 14, 16-29, 35, 37	Occasional, ususally in pairs. Generally scarce in Louidsiades where it is replaced by the closely related C. peleuensis.	6-45
C. rafflesi Bennett, 1830	1-7, 10-22, 24-29, 34-36, 39-42, 44-48, 50, 51, 54, 56, 57	Common, one of the most abundant butterflyfishes in MBP; seen on most dives.	1-15
C. reticulatus Cuvier, 1831	43	Rare, one pair seen. Photographed.	1-35
C. semeion Bleeker, 1855	3, 11, 15, 17, 19, 21, 22, 25-27, 35, 36, 40, 43-45, 52, 56, 57	Occasional.	1-25
C. speculum Cuvier, 1831*	8, 9, 11, 12, 14, 15, 18-21, 25, 26, 29-31, 34, 43, 48	Occasional.	1-30
C. trifascialis Quoy and Gaimard, 1824	4-10, 18, 19, 21, 23, 28, 31, 35, 36, 39, 40, 43, 48, 54	Occasional in areas of tabular Acropora.	2-30
C. ulietensis Cuvier, 1831	1-7, 11, 14, 16, 19, 21, 22, 25, 26, 29-31, 33, 35-40, 43-56	Moderately common.	8-30
C. unimaculatus Bloch, 1787*	4-7, 29, 37, 42, 43, 47, 48, 54, 55	Occasional.	1-60
C. vagabundus Linnaeus, 1758	1-8, 11-57	Common, the most abundant butterflyfish in MBP; several seen on almost every dive.	1-30
Chelmon rostratus (Linnaeus, 1758)	15, 36, 40-42, 46	Occasional, mainly on silty reefs of the Louisiades.	1-15
Coradion altivelis McCulloch, 1916	33, 34, 56	Rare, about 4 pairs seen on silty reefs in the Louisiades.	3-20
C. chrysozonus Cuvier, 1831	6, 9, 10, 12, 13, 15-17, 32, 40, 56	Occasional, in sheltered areas. Photographed.	5-60
C. melanopus (Cuvier, 1831)	2, 3, 7, 11, 22, 24, 26, 27	Occasional, in sheltered areas. Photographed.	10-30
Forcipger flavissimus Jordan and McGregor, 1898	4, 14, 29, 30, 37-39, 43-45, 47-50, 52-55	Occasional, mainly on outer reef slopes. Photographed.	2-114
F. longirostris (Broussonet, 1782)	29, 30, 37, 39, 43-45, 49, 50, 52-54	Occasional, mainly on outer reef slopes of the Louisiades.	5-60
Hemitaurichthys polylepis (Bleeker, 1857)*	2, 19, 39, 44, 52	Rare, but locally abundant.	3-60
Heniochus acuminatus (Linnaeus, 1758)	15, 19, 21-23, 31, 47, 56	Occasional.	2-75
H. chrysostomus Cuvier, 1831	1-57	Common, one of most abundant butterflyfishes in MBP.	5-40
H. diphreutes Jordan, 1903	2, 5, 27	Rare, only a few individuals seen.	15-210
H. monoceros Cuvier, 1831	12, 28, 52-54	Occasional pairs seen.	2-25

Milne Bay Province, Papua New Guinea—Survey II (2000) | 133

Species	Site records	Abundance	Depth (m)
H. singularius Smith and Radcliffe, 1911	4, 6-8, 10-13, 15, 16, 19-22, 25, 28, 30, 32, 34, 35, 43-48, 52, 54	Occasional, usually in clear offshore waters.	12-45
H. varius (Cuvier, 1829)	2-29, 32-35, 38-40, 42-50, 52-57	Common, one of most abundant butterflyfishes in MBP.	2-30
Parachaetodon ocellatus (Cuvier, 1831)	5, 7, 8, 10-13, 34, 46	Occasional in silty areas.	5-40
POMACANTHIDAE			
Apolemichthys trimaculatus (Lacepède, 1831)	5, 6, 19, 30, 47, 52, 53, 55	Occasional on dropoffs. Photographed.	10-50
Centropyge aurantia Randall & Wass, 1974	50	Rare, one seen at 40 m depth in rich coral area.	3-50
C. bicolor (Bloch, 1798)	1, 2, 4-10, 13, 14, 16, 17, 19, 20, 22, 28-56	Common.	3-35
C. bispinosus (Günther, 1860)	1, 2, 4, 5, 7, 14, 16-19, 29-35, 37-39, 42-45, 48-50, 52-55	Common on seaward slopes, but rare inshore. Photographed.	10-45
C. colini Smith-Vaniz & Randall, 1974	Recorded previously by B. Halstead.		25-75
C. flavicauda Fraser-Brunner, 1933	2, 6, 30, 48, 50, 55	Generally rare, but sometimes locally common on rubble bottoms.	10-60
C. loricula (Günther, 1874)	Recorded previously by B. Halstead.		5-60
C. nox (Bleeker, 1853)	5, 7-9, 12-14, 17-20, 22-28, 30-38, 41, 42, 49, 50, 54-57	Moderately common.	10-70
C. vroliki (Bleeker, 1853)	1-36, 38-57	Common, one of the two most abundant angelfishes in MBP.	3-25
Chaetodontoplus melanosoma (Bleeker, 1853)	Seen on 1997 survey only.		5-40
C. mesoleucus (Bloch, 1787)	2, 3, 7, 8, 10-15, 20, 34, 36, 40-42, 46, 56	Moderately common, but largely restricted to sheltered inshore reefs.	1-20
Genicanthus lamarck Lacepède, 1798	5, 6, 9, 16, 29, 55, 57	Occasional, but locally common; extraordinarily abundant at site 55. Photographed.	15-40
G. melanospilos (Bleeker, 1857)	2, 9, 16, 18, 24-26, 29, 30, 44, 49, 50, 52, 53	Occasional, but locally common.	20-50
Paracentropyge multifasciatus (Smith and Radcliffe, 1911)	2, 24, 25, 29, 37, 38, 44, 49, 50, 52, 53	Occasional, but seldom noticed due to cave-dwelling habits.	10-50
Pomacanthus annularis (Bloch, 1787)	Recorded previously by B. Halstead.		1-60
Pomacanthus imperator (Bloch, 1787)	6, 9, 19, 27, 30, 45, 47, 48, 52, 54-56	Occasional and in low numbers.	3-70
P. navarchus Cuvier, 1831	1, 2, 5, 6, 8, 11, 12, 14, 16-18, 20-22, 25-28, 43	Occasional.	3-30
P. semicirculatus Cuvier, 1831	10, 12, 14, 18, 22, 26, 51	Occasional.	5-40
P. sexstriatus Cuvier, 1831	4, 5, 9, 11, 14, 15, 26, 30-37, 45-52, 54-57	Occasional.	3-50
P. xanthometopon (Bleeker, 1853)	17, 30, 35, 43, 44, 52	Rarely seen, mainly on outer reef slopes.	5-30
Pygoplites diacanthus (Boddaert, 1772)	1-57	Common, one of the two most abundant angelfishes in MBP.	3-50

continued

Species	Site records	Abundance	Depth (m)
CEPOLIDAE			
Cepola sp.	Recorded previously by B. Halstead.		10-30
MUGILIDAE			
Crenimugil crenilabis (Forsskål, 1775)	8, 12, 24, 26, 27, 32	Occasional, usually in small schools containing up to about 20 fish.	0-4
Liza vaigiensis (Quoy and Gaimard, 1825)	Seen on 1997 survey only.	School of approximately 50 individuals seen at one site.	0-3
Valamugil seheli (Forsskål, 1775)	11	Rare, a school containing about 20 fish seen.	0-4
POMACENTRIDAE			
Abudefduf lorenzi Hensley and Allen, 1977	11, 12, 15, 24, 26, 32	Occasional, but locally common in shallow water next to shore. Rarely seen in the Louisiades.	0-6
A. septemfasciatus (Cuvier, 1830)	10, 32	Rare, but surge zone environment not regularly surveyed.	1-3
A. sexfasciatus Lacepède, 1802	3, 8, 9, 11, 12, 14, 18, 23, 36, 40, 44	Occasional, but sometimes locally common.	1-15
A. sordidus (Forsskål, 1775)	4, 8, 32	Rare, but surge zone environment not regularly surveyed.	1-3
A. vaigiensis (Quoy and Gaimard, 1825)	2-4, 7-9, 12, 14, 15, 18, 21-27, 30, 32, 45, 49, 51-53	Generally common.	1-12
Acanthochromis polyacantha (Bleeker, 1855)	1-33, 35-40, 43-57	Abundant in wide range of habitats. Some populations with white tails.	1-50
Amblyglyphidodon aureus (Cuvier, 1830)	2, 4-6, 9, 13, 16-20, 22, 24-29, 32, 35, 37, 43, 44, 52, 53, 55, 57	Common on steep slopes, but in low numbers.	10-35
A. batunai Allen, 1995	3-5, 12, 56	Rare, but locally common. Photographed.	2-12
A. curacao (Bloch, 1787)	1, 4-46, 48-51, 54, 56, 57	Common.	1-15
A. leucogaster (Bleeker, 1847)	1-20, 22, 24-46, 48-57	Common.	2-45
A. ternatensis (Bleeker, 1853)	8, 10, 11, 13, 15, 57	Rare, but locally common on silty inshore reefs.	2-12
Amblypomacentrus breviceps (Schlegel and Müller, 1839-44)	3, 36, 42	Rarely seen, usually, around debris and small coral outcrops situated on silt bottoms.	2-35
Amphiprion chrysopterus Cuvier, 1830	1-9, 13, 18-20, 26, 27, 29, 30, 39, 43, 44, 47, 49, 50, 52-55, 57	One of the two most common anemonefishes in MBP.	1-20
A. clarkii (Bennett, 1830)	2-4, 8, 13, 14, 16, 18-22, 24, 25, 30, 33-35, 37, 38, 42, 47, 50, 54-57	One of the two most common anemonefishes in MBP. Photographed.	1-55
A. leucokranos Allen, 1973	1, 6, 27	Rare. Photographed.	2-12
A. melanopus Bleeker, 1852	2, 4, 6, 9, 19-21, 26, 27, 29, 34, 36, 37, 43-45, 48, 53, 55-57	Moderately common.	1-10
A. percula (Lacepède, 1802)	4, 6, 12, 16, 20, 22, 26, 27, 32, 36, 42, 56	Moderately common.	1-15
A. perideraion Bleeker, 1855	3, 4, 6, 7, 9, 14, 16, 19, 20, 26, 30, 31, 37, 39, 43, 52, 55, 57	Moderately common. Photographed.	3-20

continued

Species	Site records	Abundance	Depth (m)
A. polymnus (Linnaeus, 1758)	23	Rare, but restricted to featureless silt or sand bottoms away from reefs.	2-30
A. sandaracinos Allen, 1972	2, 9, 14, 20	Rare, only about 10 seen.	3-20
Cheiloprion labiatus (Day, 1877)	56	Rare.	1-3
Chromis alpha Randall, 1988	17, 20, 25-30, 43, 44, 49, 50, 52-54	Moderately common on steep slopes. Photographed.	18-95
C. amboinensis (Bleeker, 1873)	1-9, 12-20, 22-39, 42-45, 48-50, 52-55, 57	Abundant. Photographed.	5-65
C. analis (Cuvier, 1830)	1, 2, 5, 17-20, 22, 24-28, 30, 44, 52, 53, 57	Moderately common on steep slopes.	10-70
C. atripectoralis Welander and Schultz, 1951	3-5, 9, 12, 14, 16-29, 37, 43, 48, 54	Abundant.	2-15
C. atripes Fowler and Bean, 1928	1, 2, 4-9, 12-14, 18, 19, 22, 24-30, 33, 34, 37, 39, 43-45, 48-50, 52-55	Common, particularly on slopes.	10-35
C. caudalis Randall, 1988	Seen on 1997 survey only.		20-50
C. delta Randall, 1988	1, 2, 4, 6, 9, 17-19, 22-30, 33, 37, 38, 43, 44, 49, 50, 52, 53, 55	Common, especially on steep slopes below about 15 m depth.	10-80
C. elerae Fowler and Bean, 1928	4, 5, 9, 13, 19, 22, 24, 25, 27, 28, 37, 38, 44	Common, always in caves and crevices on steep slopes.	12-70
C. iomelas Jordan & Seale, 1906*	49, 52, 53	Rare, but locally common at three sites in W. Louisiades.	5-40
C. lepidolepis Bleeker, 1877	1, 2, 4, 6-9, 14, 18-20, 23, 25, 26, 28-32, 34, 35, 37-39, 42-45, 48-50, 54, 55	Common. Photographed.	2-20
C. lineata Fowler and Bean, 1928	1, 2, 8, 14, 17, 18, 20, 22-28, 30, 35, 37, 43-45, 47-50, 52, 53	Common, usually in clear water with some wave action.	2-10
C. margaritifer Fowler, 1946	1-10, 12-14, 17-32, 35, 37-39, 42-45, 47-50, 52-55	Common in clear water areas.	2-20
C. retrofasciata Weber, 1913	1-46, 48-51, 54-57	Abundant at most sites.	5-65
C. ternatensis (Bleeker, 1856)	1-10, 12-14, 16-40, 43-51, 53-57	Abundant, often forming dense shoals on the edge of steep slopes.	2-15
C. viridis (Cuvier, 1830)	3, 4, 7-15, 18, 21-26, 36, 38-40, 42, 43, 46, 48, 51, 56	Abundant in sheltered areas of rich coral, generally in clear water.	1-12
C. weberi Fowler and Bean, 1928	1, 2, 4-7, 9, 18-20, 26, 28, 30-35, 39, 43, 45, 47-50, 53-55	Common.	3-25
C. xanthochira (Bleeker, 1851)	1, 2, 4-6, 17, 19, 26, 28-30, 35, 39, 43, 44, 48-50, 52-55	Common on outer slopes.	10-48
C. xanthura (Bleeker, 1854)	2-7, 12-14, 16-20, 22-40, 43-45, 48-50, 52, 53	Common, especially on steep slopes.	3-40
Chrysiptera biocellata (Quoy and Gaimard, 1824)	7, 11, 35, 39, 51	Generally rare, but locally common in well-sheltered sandy areas with coral outcrops next to shore.	0-5
C. caeruleolineata (Allen, 1973)	10, 16, 30, 57	Probably common, but deep habitat not properly surveyed. Photographed.	30-65
C. cyanea (Quoy and Gaimard, 1824)	3, 4, 11, 12, 14, 15, 18, 21, 22, 26, 35, 37, 38, 40, 41, 42, 45, 48, 49, 51, 56, 57	Moderately common, usually in shallow well-sheltered areas with clear water.	0-10

continued

continued

Species	Site records	Abundance	Depth (m)
C. cymatilis Allen, 1999	3, 7, 8, 10-15, 20, 26, 27, 32, 34-36, 40, 42, 46, 51, 56, 57	Common in sheltered silty areas with minimal currents. Photographed Endemic to Milne Bay Province.	4-45
C. flavipinnis (Allen and Robertson, 1974)	1, 5, 6, 17, 19, 28-31, 44, 48-50, 52, 54, 55	Occasional, always in sand or in rubble areas.	5-40
C. leucopoma (Lesson, 1830)	4, 8, 14, 15, 18, 30, 32, 33, 45, 45, 47, 49	Moderately common, usually in shallow beach rock areas affected by surge.	0-2
C. oxycephala (Bleeker, 1877)	7, 8, 11-13, 15, 36, 40-42, 46, 51, 56, 57	Moderately common, usually on silty inshore reefs with abundant coral. Photographed.	1-16
C. rex (Snyder, 1909)	4, 14, 18, 20, 22, 24, 29, 30, 31, 37-39, 43-45, 47-50	Occasional, usually in areas with some surge.	1-6
C. rollandi (Whitley, 1961)	1-29, 31-43, 46, 51, 56, 57	Moderately common, particularly on reef slopes affected by silt.	2-35
C. talboti (Allen, 1975)	1, 2, 4-9, 12-14, 16-35, 37-39, 42-45, 48-50, 52-55, 57	Moderately common, except in silty areas.	6-35
C. unimaculata (Cuvier, 1830)	12, 15, 56	Occasional, but locally common.	0-2
Dascyllus aruanus (Linnaeus, 1758)	3, 7, 10-15, 17, 21-23, 26, 32-35, 37-43, 46, 48, 51, 54, 56, 57	Common, forming aggregations around small coral heads.	1-12
D. melanurus Bleeker, 1854	3, 7, 10-12, 14, 15, 21-23, 26, 33, 35, 42, 51, 57	Occasional.	1-10
D. reticulatus (Richardson, 1846)	1-9, 12-14, 16-33, 35, 37-40, 43-56	Common.	1-50
D. trimaculatus (Rüppell, 1928)	2-9, 14, 16-20, 22-39, 43-45, 47-50, 53-56	Common.	1-55
Dischistodus chrysopoecilus (Schlegel and Müller, 1839)	7, 12, 38, 57	Generally rare, but locally common in sand-rubble areas near shallow seagrass beds.	1-5
D. melanotus (Bleeker, 1858)	3, 4, 7, 8, 10-15, 17, 21, 22, 26-28, 32, 33, 35, 37, 38, 40-42, 51, 54, 56, 57	Occasional.	1-10
D. perspicillatus (Cuvier, 1830)	3, 7, 10, 11, 32, 40, 51, 54, 57	Occasional.	1-10
D. prosopotaenia (Bleeker, 1852)	7, 10, 11, 14, 15, 21, 36, 38, 40-42, 46, 51, 56, 57	Occasional. Photographed.	1-12
D. pseudochrysopoecilus Allen and Robertson, 1974	4, 7, 10, 12, 15, 22, 29, 33, 35, 37, 45, 49, 51, 54, 57	Occasional.	1-5
Hemiglyphidodon plagiometopon (Bleeker, 1852)	3, 8, 10-13, 15, 21, 23, 36, 40, 42, 46, 51, 56, 57	Moderately common, generally on sheltered reefs affected by silt.	1-20
Lepidozygus tapeinosoma (Bleeker, 1856)	48, 52	Rare, mainly in clean water with strong currents.	5-25
Neoglyphidodon melas (Cuvier, 1830)	4, 6, 10-13, 15, 21, 22, 27, 33, 36, 37, 42, 48, 49, 54, 56, 57	Moderately common, but in low numbers at each site.	1-12
N. nigroris (Cuvier, 1830)	1-47, 48-50, 54-57	Common.	2-23
N. thoracotaeniatus (Fowler and Bean, 1928)	5-9, 12-14, 16-18, 20, 22, 24-30, 37, 49, 50, 57	Moderately common, usually on sheltered slopes. Less common in the Louisiades. Photographed.	15-45
Neopomacentrus aquadulcis Jenkins and Allen, 2000	Recorded in 1999 by G. Allen.		0-3

Species	Site records	Abundance	Depth (m)
N. azysron (Bleeker, 1877)	14, 15, 31, 36, 56	Occasional, but locally common at some sites.	1-12
N. cyanomos (Bleeker, 1856)	34	Rare.	5-18
N. filamentosus (Macleay, 1833)	15, 40, 41, 42, 46	Occasional, but locally common.	5-15
N. nemurus (Bleeker, 1857)	22-26	Occasional, but locally common on sheltered inshore reefs.	1-10
N. violascens (Bleeker, 1848)	Seen on 1997 survey only.		5-25
Plectroglyphidodon dickii (Liénard, 1839)	4, 8, 18-20, 25, 27, 39, 43-45, 52, 53	Occasional.	1-12
P. johnstonianus Fowler & Ball, 1924	52	Rare, several seen at only one site.	2-12
P. lacrymatus (Quoy and Gaimard, 1824)	1, 3, 4, 6-12, 14, 15, 17-33, 35-57	Common.	2-12
P. leucozonus (Bleeker, 1859)	4, 8, 18, 24, 52	Occasional.	0-2
Pomacentrus adelus Allen, 1991	2, 4, 7, 8, 10-15, 17-22, 24-28, 31-33, 35-57	Common.	0-5
P. amboinensis Bleeker, 1868	1, 3-39, 41-46, 48-51, 54, 55, 57	Abundant.	2-40
P. bankanensis Bleeker, 1853	1-12, 14-31, 33, 36, 37, 39, 41-50, 52-55, 57	Common.	0-12
P. brachialis Cuvier, 1830	1-5, 7-9, 12-14, 16-22, 24-26, 29-41, 43, 44, 47-51, 54, 55, 57	Abundant, especially in areas exposed to curents.	6-40
P. burroughi Fowler, 1918	3, 7, 8, 10-15, 21-23, 26, 35, 36, 40-42, 46, 51, 56	Moderately common, usually on silty inshore reefs.	2-16
P. chrysurus Cuvier, 1830	4, 7, 8, 10-12, 15, 32, 37, 39-41, 49, 51, 56, 57	Occasional, around small coral or rock formations surrounded by sand.	0-3
P. coelestis Jordan and Starks, 1901	1-8, 10-12, 14, 16-20, 22, 23, 26, 28-35, 38-41, 43-45, 47-50, 53-55	Common.	1-12
P. colini Allen, 1991	15, 36	Rare, but locally common in strongly silted areas.	20-28
P. grammorhynchus Fowler, 1918	3, 7, 8, 10-12, 15, 21, 22, 35, 36, 40, 41, 42, 46, 51, 54, 56, 57	Occasional, but locally common among live and dead corals (often staghorn Acropora). Photographed.	2-12
P. lepidogenys Fowler and Bean, 1928	1-12, 14, 15, 17-37, 42-45, 48-50, 52-56	Common.	1-12
P. moluccensis Bleeker, 1853	1-57	Abundant.	1-14
P. nagasakiensis Tanaka, 1917	9-11, 13, 17, 18, 31-33, 35, 38, 40-42, 46, 56	Occasional, around isolated rocky outcrops surrounded by sand.	5-30
P. nigromanus Weber, 1913	1, 3-8, 10-28, 31-43, 46, 51, 56, 57	Common, usually on slopes in a variety of habitats.	6-60
P. nigromarginatus Allen, 1973	1, 5-8, 10-14, 16-20, 22, 24-29, 30, 37, 44, 45, 48-50, 52, 53, 57	Moderately common on steep slopes.	20-50
P. opisthostigma Fowler, 1918	7, 8, 10, 11, 13, 15	Locally common in silty habitats, mainly in the Amphlett and D'Entrecasteaux Islands.	10-25
P. pavo (Bloch, 1878)	3, 7, 10, 11, 21, 42, 46, 48, 51	Moderately common, always around coral patches in sandy lagoons.	1-16

continued

Species	Site records	Abundance	Depth (m)
P. philippinus Evermann and Seale, 1907	4-7, 12, 13, 17, 18, 24, 25, 27, 29, 30, 31, 36-39, 42-45, 48-50, 52-54	Moderately common. Photographed.	1-12
P. reidi Fowler and Bean, 1928	1, 2, 4-9, 12, 13, 16-20- 22-27, 29-31, 33, 37, 38, 43-45, 49, 50, 52, 53, 55	Moderately common, usually on seaward slopes.	12-70
P. simsiang Bleeker, 1856	3, 11, 12, 15, 21-23, 46, 51, 57	Moderately common, usually in areas with silty bottoms.	0-10
P. smithi Fowler and Bean, 1928	3, 7, 8, 10-13, 15, 21, 26, 27, 35, 36, 38, 40-42, 46, 51, 56, 57	Abundant on silty coastal reefs, especially at sites 42 and 56.	2-14
P. taeniometopon Bleeker, 1852	Recorded in 1999 by GRA.		0-3
P. tripunctatus Cuvier, 1830	Seen on 1997 survey only.		0-3
P. vaiuli Jordan and Seale, 1906	30, 44, 45, 49, 50, 52	Occasional at outer reef sites in the Louisiades. Photographed	3-45
Premnas biaculeatus (Bloch, 1790)	1-5, 9-12, 14-16, 27, 29, 30, 32, 33, 35, 38, 42, 48-50, 54, 56, 57	Occasional. Photographed.	1-6
Stegastes albifasciatus (Schlegel and Müller, 1839)	12, 22, 32, 51, 53	Occasional, but sometimes locally common.	0-2
S. fasciolatus (Ogilby, 1889)	4, 14, 18, 24, 25, 29-31, 43-45, 49, 50, 52, 53	Occasional, but locally common. Photographed.	0-5
S. lividus (Bloch and Schneider, 1801)	3, 4, 11, 12, 22, 51, 57	Occasional, but locally common.	1-5
S. nigricans (Lacepède, 1802)	3, 4, 7, 8, 10, 11, 15, 21, 22, 29, 32, 33, 35-37, 39, 40, 44-46, 48, 49, 51, 54, 56	Occasional, but locally common.	1-12
LABRIDAE			
Anampses caeruleopunctatus Rüppell, 1828	8	Rare, one female seen.	2-30
A. geographicus Valenciennes, 1840	29, 48	Rare, only 3 individuals seen.	5-25
A. melanurus Bleeker, 1857	52	Rare, only one seen.	12-40
A. meleagrides Valenciennes, 1840	5, 6, 18, 19, 29, 47, 48, 54	Occasional, always in small numbers.	4-60
A. neoguinaicus Bleeker, 1878	1, 3-7, 9, 14, 16-20, 26, 27, 29-31, 33, 35, 37, 39, 43-45, 47-50, 52-55	Moderately common, several seen on most dives.	8-30
A. twistii Bleeker, 1856	Seen on 1997 survey only.		2-30
Bodianus anthioides (Bennett, 1831)	2, 19, 22, 26, 44, 49, 52, 53	Rare, less than 10 seen.	6-60
B. axillaris (Bennett, 1831)	39, 44, 45, 47, 49, 52-54	Rare, about 10 seen.	2-40
B. bimaculatus Allen, 1973	53	Rare, several seen below 45 m.	30-60
B. diana (Lacepède, 1802)	1, 2, 4-6, 9, 13, 16, 18-20, 22-32, 35, 37, 39, 43-45, 47-50, 52-55, 57	Moderately common. Photographed.	6-25
B. izuensis Araga & Yoshino, 1975	Recorded previously by B. Halstead.		40-100

continued

Species	Site records	Abundance	Depth (m)
B. loxozonus (Snyder, 1908)	38, 39, 43, 45, 52, 54	Rare, less than 10 seen.	3-40
B. mesothorax (Bloch & Schneider, 1801)	1-40, 42-45, 48-50, 52-57	Common.	5-30
B. sp. 1	Recorded previously by B. Halstead.		30-70
B. sp. 2	Recorded previously by B. Halstead.		30-70
Cheilinus chlorurus (Bloch, 1791)	31, 40-43, 46, 51, 55, 56	Occasional.	2-30
C. fasciatus (Bloch, 1791)	1-32, 34-45, 47-57	Moderately common, severaladults seen on most dives.	4-40
C. oxycephalus (Bleeker, 1853)	3, 4, 6-9, 12-14, 17, 27, 29, 33, 35, 39-46, 48-52, 54-56	Moderately common.	1-20
C. trilobatus Lacepède, 1801	4, 6, 7, 10-15, 18-20, 22, 30, 44, 45, 47-50, 52, 54-57	Moderately common, severaladults seen on most dives.	1-20
C. undulatus Rüppell, 1835	4, 6, 8, 9, 11, 12, 15-20, 30, 31, 34-37, 39, 43-49, 52, 57	Moderately common, but always in small numbers.	2-60
Cheilio inermis (Forsskål, 1775)	12, 14, 45	Rare, but mostly in weed habitats.	0-3
Choerodon anchorago (Bloch, 1791)	3, 4, 8, 10-12, 15, 18, 20-23, 33-35, 38, 40, 41, 46, 56, 57	Occasional, usually in silty areas.	1-25
C. fasciatus (Günther, 1867)*	34, 49	Rare, less than 10 seen.	3-20
C. jordani (Snyder, 1908)	Seen on 1997 survey only.		10-20
C. zosterophorus (Bleeker, 1868)	4, 9, 14, 16, 33, 35, 51, 56, 57	Occasional in small groups over sand bottoms. Photographed.	5-50
Cirrhilabrus condei Allen and Randall, 1996	1, 6, 54	Rare. Previously recorded usually form depth below 20-30 m, but seen as shallow as 5 m at site 54.	25-45
C. exquisitus Smith, 1957	6, 19, 31, 45, 47, 54	Occasional.	6-32
C. punctatus Randall and Kuiter, 1989	1-9, 11-57	Abundant, one of most common labrids in MBP. Photographed.	3-60
C. pylei Allen & Randall, 1996	Recorded previously by B. Halstead.	Endemic to Milne Bay Province.	60-90
C. walindi Allen, 1995	2, 20, 24-26, 28	Rare, several individuals seen in 35-40 m, but survey did not adequately sample its normal depth range, which lies below 30-40 m. Photoghraphed.	20-60
C. sp.*	40, 42, 44	Rare, uusually seen with Ciirhilabrus punctatus.	5-30
Coris aygula Lacepède, 1802	43, 54	Rare, only three adults seen.	1-50
C. batuensis (Bleeker, 1862)	1, 3, 8, 10, 11, 14-16, 19-22, 26, 29-39, 44, 46, 48-51, 54, 56, 57	Moderately common.	3-25
C. caudimacula (Quoy and Gaimard, 1834)	Seen on 1997 survey only.		3-50
C. dorsomacula Fowler, 1908	6, 29, 45, 47	Rare. Photographed.	4-25
C. gaimardi (Quoy and Gaimard, 1824)	4-7, 10, 14, 17, 19, 20, 29, 31, 32, 39, 44, 45, 50, 54	Occasional.	1-50
Cymolutes torquatus Valenciennes, 1840	Seen on 1997 survey only.		3-20

continued

continued

Species	Site records	Abundance	Depth (m)
Diproctacanthus xanthurus (Bleeker, 1856)	1, 3-5, 7-17, 20, 26, 28, 31, 32, 34-36, 40-42, 46, 51, 55-57	Moderately common, but most abundant on protected inshore reefs.	2-15
Epibulus insidiator (Pallas, 1770)	1-9, 12, 14-45, 47-57	Common. Photographed.	1-40
Gomphosus varius Lacepède, 1801	1-10, 12, 14-44, 47-57	Common.	1-30
Halichoeres argus (Bloch and Schneider, 1801)	14, 15, 22, 27, 38, 40-42	Occasional, usually in silty protected areas with weeds.	0-3
H. binotopsis (Bleeker, 1849)	26, 27	Rare, about five seen.	2-20
H. biocellatus Schultz, 1960	1, 4, 5, 7, 9, 16, 17, 19, 24, 29, 30, 39, 43-45, 48-50, 52, 53, 55	Occasional. Photographed.	6-35
H.chloropterus (Bloch, 1791)	3, 4, 7, 10-12, 14, 15, 20-23, 26, 32, 33, 35, 36, 38, 40-42, 46, 51, 56, 57	Moderately common, ususally on protected inshore reefs with sand and weeds.	0-10
H.chrysus Randall, 1980	1, 4-7, 9, 16, 18-20, 22, 28-31, 39, 43-45, 47-50, 53-55	Moderately common on clean sand bottoms.	7-60
H. dussumieri (Valenciennes, 1839)	Recorded previously by B. Halstead.		0-10
H. hartzfeldi Bleeker, 1852	1, 31	Rare.	10-30
H. hortulanus (Lacepède, 1802)	1, 3-10, 12-15, 17-22, 25-33, 35, 37, 39-45, 47-55	Common.	1-30
H. leucurus (Walbaum, 1792)	3, 8, 10-13, 15, 20, 23, 26, 27, 36, 40, 41, 46, 56, 57	Occasional on silty inshore reefs.	1-20
H. margaritaceus (Valenciennes, 1839)	4, 7, 10, 12, 14, 15, 18, 20, 21, 29, 31, 33, 35, 39, 45, 47-49, 51, 54, 55	Modertely common, usually at sites that included shallow water next to shore.	0-3
H. marginatus (Rüppell, 1835)	4, 7, 8, 10, 12, 14, 15, 17, 18, 21, 30, 31, 33-35, 40, 42, 43, 47, 51, 53, 55	Moderately common.	1-30
H. melanurus (Bleeker, 1851)	3, 4-20, 22-24, 26, 31-33, 35-37, 40-42, 46, 51, 55-57	Common.	2-15
H. miniatus (Valenciennes, 1839)	4, 8, 10	Rare, but locally common.	0-8
H. ornatissimus (Garrett, 1863)	6, 7, 43, 44, 49, 50	Rare.	5-25
H. prosopeion (Bleeker, 1853)	1, 4-9, 12-14, 16-20, 22-41, 43-45, 48-50, 52-55, 57	Common in variety of habitats. Photographed.	5-40
H. scapularis (Bennett, 1832)	3, 4, 20, 21, 39, 57	Occasional, always in sandy areas.	0-15
H. trimaculatus (Quoy & Gaimard, 1834)	3, 4, 7, 10-12, 18, 20, 29, 32, 37-39, 43-45, 47, 48, 50, 51, 54, 57	Moderately common, usually in sandy areas.	0-20
Hemigymnus fasciatus (Bloch, 1792)	1-12, 14-22, 24-45, 48-57	Common, but in low numbers at each site.	1-20
H. melapterus (Bloch, 1791)	3-8, 10-22, 26-57	Common, but in low numbers at each site.	2-30
Hologymnosus annulatus (Lacepède, 1801)	Seen on 1997 survey only.		5-30
H. doliatus (Lacepède, 1801)	5, 6, 8, 19-21, 26, 43-45, 47, 48, 50, 55, 56	Moderately common, but mainly juveniles seen.	4-35

Species	Site records	Abundance	Depth (m)
Labrichthys unilineatus (Guichenot, 1847)	1-29, 35, 37, 42-44, 46, 48, 54-57	Common, especially in rich coral areas.	1-20
Labroides bicolor Fowler and Bean, 1928	2-4, 8, 9, 12, 14, 18, 19, 22-27, 29-31, 35, 39, 40, 42-45, 47-50, 52, 54, 55	Occasional, generally in much smaller numbers than other *Labroides* species.	2-40
L. dimidiatus (Valenciennes, 1839)	1-14, 16-57	Moderately common.	1-40
L. pectoralis Randall and Springer, 1975	2, 4, 13, 14, 17-22, 24-26, 29, 43-45, 48, 50	Occasional.	2-28
Labropsis alleni Randall, 1981	2, 22, 24-26, 44	Rare, but moderately common at several sites in Goodenough Bay.	4-52
L. australis Randall, 1981	2, 19, 25, 29, 30, 33, 35, 38, 48, 49, 54, 55	Occasional.	2-55
L. xanthonota Randall, 1981	43, 44, 48-50	Generally rare.	1-30
Leptojulis urostigma Randall, 1996	33	Rare, but easily overlooked due to sandy habitat.	15-80
Macropharyngodon meleagris (Valenciennes, 1839)	4, 10, 12, 14, 18, 27, 29-31, 35, 39, 43-45, 47-50, 52-55	Moderately common, but always in small numbers at each site.	1-30
M. negrosensis Herre, 1932	2, 5, 31, 32, 44, 47, 48, 55	Occasional.	8-30
Novaculichthys macrolepidotus (Bloch, 1791)	Recorded previously by B. Halstead.		1-12
N. sp.	Recorded previously by B. Halstead.	A new species collected by J. Randall. Endemic to Milne Bay Province.	20-50
N. taeniourus (Lacepède, 1802)	10, 33, 39, 40, 42, 44, 45, 54	Occasional.	1-14
Oxycheilinus arenatus (Valenciennes, 1840)	44	One seen in 50 m depth.	20-80
O. bimaculatus (Valenciennes, 1840)	1, 8, 35, 42, 44, 46	Occasional, around rock and coral outcrops on sandy or rubble bottoms.	2-110
O. celebicus (Bleeker, 1853)	3-5, 7, 8, 10-17, 20, 22-28, 35, 36, 40-42, 46, 51, 56, 57	Occasional on sheltered inshore reefs. Photographed.	3-30
O. diagrammus (Lacepède, 1802)	1-4, 6, 7, 9, 10, 12, 15-20, 22, 25, 26, 29, 30, 32, 33, 35, 37, 39, 42-45, 48-55	Occasional.	3-120
O. orientalis (Günther, 1862)	2-5, 7, 13, 14, 26, 27, 30, 32, 35, 44, 45, 48, 52-54	Occasional.	15-70
O. unifasciatus (Streets, 1877)	Recorded previously by B. Halstead.		3-80
Parachelinus filamentosus Allen, 1974	1, 2, 3-23, 27-30, 33, 37, 38, 41, 42, 46, 49, 50, 53, 54, 56	Common, usually in rubble areas.	10-50
Pseudocheilinops ataenia Schultz, 1960*	13, 41	Rare, only two fish seen.	5-25
Pseudocheilinus evanidus Jordan and Evermann, 1902	2, 4-10, 13, 16-21, 23, 25, 26, 28-31, 37-39, 43-45, 48-50, 52-55	Occasional.	6-40
P. hexataenia (Bleeker, 1857)	2-11, 14-20, 22, 24-31, 34, 37, 39, 42-45, 47-50, 52-55	Moderately common, only a few seen on each dive, but has cryptic habits.	2-35
Pseudocoris heteroptera (Bleeker, 1857)	19, 22, 53	Rare, only one male and three females seen.	10-30
P. yamashiroi (Schmidt, 1930)	5, 19, 48, 49, 53, 55	Occasional.	10-30

continued

Species	Site records	Abundance	Depth (m)
Pseudodax moluccanus (Valenciennes, 1840)	Seen on 1997 survey only.		3-40
Pteragogus cryptus Randall, 1981	16, 46, 51	Rare, but has cryptic habits.	4-65
P. enneacanthus (Bleeker, 1856)	Recorded previously by B. Halstead.		5-40
Stethojulis bandanensis (Bleeker, 1851)	4, 6, 8, 10, 12, 14, 15, 29, 33, 35,37, 43, 46, 51, 54	Moderately common.	0-30
S. interrupta (Bleeker, 1851)	3, 13-15, 18, 29, 32, 33, 42, 46, 48, 51, 57	Moderately common. Photographed.	4-25
S. strigiventer (Bennett, 1832)	Seen on 1997 survey only.		0-6
S. trilineata (Bloch and Schneider, 1801)	3, 4, 8-10, 12, 14, 15, 20, 21, 22, 24, 29-31, 33, 37-40, 42-46, 48-51	Moderately common.	1-10
Thalassoma amblycephalum (Bleeker, 1856)	1-14, 17-22, 24-33, 35, 37-39, 43-45, 47-50, 52-55	Common.	1-15
T. hardwicke (Bennett, 1828)	1, 3-15, 17-22, 24-33, 35-56	Common.	0-15
T. jansenii (Bleeker, 1856)	4, 8, 14, 15, 18, 20, 24-26, 29-32, 35-37, 39, 43-45, 47-55	Moderately common, usually in very shallow water exposed to surge.	0-15
T. lunare (Linnaeus, 1758)	1-57	Common, one of most abundant wrasses.	1-30
T. lutescens (Lay & Bennett, 1839)	47, 52, 54	Rare, about five adult males seen.	1-30
T. purpureum (Forsskål, 1775)	Seen on 1997 survey only.		2-20
T. quinquevittatum (Lay and Bennett, 1839)	4, 43-45, 47, 52, 53	Rare, except locally common at a few sites exposed to surge.	0-18
T. trilobatum (Lacepède, 1801)	Seen on 1997 survey only.		0-5
Wetmorella albofasciata Schultz & Marshall, 1954*	22	Collected with rotenone.	5-40
W. nigropinnata (Seale, 1901)	53	Collected with rotenone.	5-50
Xyrichtys aneitensis (Günther, 1862)	10	Rare, only one seen on flat sand bottom.	2-25
X. pavo Valenciennes, 1839	Seen on 1997 survey only.		5-80
X. pentadactylus (Linnaeus, 1758)	Recorded previously by B. Halstead.		30-50
X. sp.	Recorded previously by B. Halstead.		
SCARIDAE			
Bolbometopon muricatum (Valenciennes, 1840)	20, 22, 35, 39, 45	Occasional, in groups of up to about 5-15 large adults.	1-30
Calotomus carolinus (Valenciennes, 1839)	49	Rare, only one adult male seen.	4-30
Cetoscarus bicolor (Rüppell, 1828)	1-4, 6-22, 25, 27-33, 35, 37-54, 57	Common, but usually in small numbers.	1-30
Chlorurus bleekeri (de Beaufort, 1940)	1, 3-7, 9-38, 49-46, 48-51, 55-57	Common, one of most abundant parrotfishes in MBP.	2-30
C. microrhinos (Bleeker, 1854)	2, 4-6, 8-22, 24-26, 28-32, 34, 35, 39, 40, 42-45, 47-50, 52-54, 56	Common. Photographed.	2-35

continued

Species	Site records	Abundance	Depth (m)
C. sordidus (Forsskål, 1775)	1-11, 13-22, 24-26, 28-37, 39-57	Common, one of most abundant parrotfishes in MBP.	1-25
Hipposcarus longiceps (Bleeker, 1862)	1-3, 6-10, 12, 13, 15, 20-22, 25, 27-29, 31, 36, 39, 40, 43-48, 51, 53, 54, 57	Common at sites adjacent to sandy bottoms.	5-40
Leptoscarus vaigiensis (Quoy & Gaimard, 1824)	Recorded previously by B. Halstead.		1-20
Scarus altipinnis (Steindachner, 1879)	6, 10, 14, 15, 18, 21, 27, 28, 31, 34, 37, 39, 43-46, 52, 57.	Occasional.	5-20
S. chameleon Choat and Randall, 1986)	4, 8, 9, 12-14, 18, 29, 30, 31, 35, 43-45, 47-50, 52-54	Occasional, always in small numbers.	3-15
S. dimidiatus Bleeker, 1859	1-7, 9-37, 39-50, 52-54, 56, 57	Moderately common.	1-15
S. flavipectoralis Schultz, 1958	1-46, 48-57	Common, one of most abundant parrotfishes in MBP. Photographed.	8-40
S. festivus Valenciennes, 1840*	27	Rare, one adult male seen.	5-30
S. forsteni (Bleeker, 1861)	1, 2, 5, 6, 9, 11, 14, 25, 27, 39, 44, 45, 47, 48, 55	Occasional, but locally common at a few sites.	3-30
S. frenatus Lacepède, 1802	6, 8, 10, 13, 15, 17, 43, 45, 47, 48, 51, 56	Occasional.	3-25
S. ghobban Forsskål, 1775	1, 3, 4, 8-10, 14, 15, 17, 21, 22, 28, 46, 57	Occasional.	3-30
S. globiceps Valenciennes, 1840	10, 15, 29-31, 34, 44, 47-50, 52, 53	Occasional.	2-15
S. niger Forsskål, 1775	1-7, 9-56	Common.	2-20
S. oviceps Valenciennes, 1839	3-15, 17, 18, 21, 22, 27, 30, 31, 37, 39, 43, 44, 46-54, 57	Moderately common.	1-12
S. psittacus Forsskål, 1775	1, 2, 10, 13, 18, 20, 29, 39, 43, 47, 48, 52-54	Occasional.	4-25
S. pyrrhurus (Jordan and Seale, 1906)	15, 24, 25, 27, 35, 45	Rare, about 10 seen.	3-20
S. quoyi Valenciennes, 1840	3-7, 10-12, 14, 15, 17, 20, 21, 26-28, 36, 38, 40, 41, 57	Moderately common, usually on protected inshore reefs with increased turbidity.	4-18
S. rivulatus Valenciennes, 1840	3, 4, 14, 15, 21, 28, 31, 41, 42, 45, 46	Occasional.	5-20
S. rubroviolaceus Bleeker, 1849	6, 10, 18-20, 26, 29-31, 39, 43-45, 47-50, 52-55	Occasional.	1-30
S. schlegeli (Bleeker, 1861)	2, 3, 6, 9,10, 29, 30, 35, 40, 45, 47, 54	Occasional.	1-45
S. sp.	19	Rare, only one adult male seen in 40 m depth.	18-60
S. spinus (Kner, 1868)	2, 6-11, 13, 14, 16-21, 24-31, 34, 35, 37, 43-45, 48-50, 52, 54, 55	Occasional. Photographed.	2-18
S. tricolor Bleeker, 1849	5	Rare, an adult pair seen in 25 m depth.	8-40
AMMODYTIDAE			
Ammodytoides sp.	Recorded previously by B. Halstead.		5-20

continued

continued

Species	Depth (m)	Abundance	Site records
TRICHONOTIDAE			
Trichonotus elegans Shimada & Yoshino, 1984	5-25		Recorded previously by B. Halstead.
T. halstead Clark and Pohle, 1996	8-25	Endemic to Milne Bay Province. Known only from Observation Pt., Fergusson Island.	Seen on 1997 survey only.
T. setiger Bloch and Schneider, 1801	3-15		Seen on 1997 survey only.
CREEDIIDAE			
Limnichthys fasciatus Waite, 1904	1-15		Recorded previously by B. Halstead.
PINGUIPEDIDAE			
Parapercis clathrata Ogilby, 1911	3-50	Occasional, the most common grubfish in MBP.	1, 4, 6, 10, 11, 14, 19, 20, 26, 29-34, 39, 43-45, 47-50, 54, 55
P. cylindrica (Bloch, 1792)	0-20	Occasional in weed-sand areas.	32, 33, 46, 54
P. hexophthalma (Cuvier, 1829)	5-25	Occasional in Louisiades.	29, 39, 43-47, 49, 54
P. millepunctata (Günther, 1860)	3-50		Seen on 1997 survey only.
P. schauinslandi (Steindachner, 1900)	15-80	Rare, only one seen in 45 m depth.	39
P. sp. 1	5-25		Seen on 1997 survey only.
P. sp. 2	5-25	Occasional, found on clean white sand.	35, 36, 42, 46
P. tetracantha (Lacepède, 1800)	8-40	Rare, only about six seen.	1, 5, 47, 55
P. xanthozona (Bleeker, 1849)	1-15	Occasional. Photographed.	15, 17, 22, 23, 36, 46
PHOLIDICHTHYIDAE			
Pholidichthys leucotaenia Bleeker, 1856	1-40	Occasional, locally common but usually only juveniles seen.	5, 8-10, 13, 14, 19, 25, 31, 43, 57
TRIPTERYGIIDAE			
Enneapterygius mirabilis Fricke, 1994	8-37	One specimen collected with rotenone.	22
E. tutuilae Jordan and Seale, 1906	0-32	Several collected with rotenone.	15, 53
Helcogramma gymnauchen (Weber, 1909)	3-15		Seen on 1997 survey only.
H. striata Hansen, 1986	1-20	Rare.	6, 33, 52, 55
Ucla xenogrammus Holleman, 1993	2-40	Rare.	4, 26
BLENNIIDAE			
Aspidontus dussumieri (Valenciennes, 1836)	1-25		Seen on 1997 survey only.
A. taeniatus Quoy & Gaimard, 1834	1-25		Recorded previously by B. Halstead.
Atrosalarias fuscus (Rüppell, 1835)	1-12	Occasional in rich coral areas.	3, 8, 10, 12, 14, 20-22, 35, 46, 48, 56, 57
Blenniella chrysospilos (Bleeker, 1857)	0-3	Rare, but not readily observed due to shallow wave-swept habitat.	8, 49

Species	Site records	Abundance	Depth (m)
B. edentulus Bloch and Schneider, 1801	Seen on 1997 survey only.		0-2
B. lineatus (Valenciennes, 1836)	Seen on 1997 survey only.		0-2
Cirripectes castaneus Valenciennes, 1836	17, 22, 24, 35, 47	Occasional.	1-5
C. filamentosus (Alleyne & Macleay, 1877)*	8, 13-15, 18, 21, 26-28, 30, 33, 40, 42, 43, 45, 49, 50, 53, 57	Moderately common.	1-20
C. polyzona (Bleeker, 1868)	Seen on 1997 survey only.		0-3
C. springeri Williams, 1988	Seen on 1997 survey only.		1-18
C. stigmaticus Strasburg and Schultz, 1953	1, 3, 4, 8, 10, 17, 18, 20, 22, 29, 30	Occasional.	0-5
Crossosalarias macrospilus Smith-Vaniz and Springer, 1971	4, 52	Rare, only two seen.	1-25
Escenius aequalis Springer, 1988	4-6, 10, 22, 24, 47	Occasional. Photographed.	2-10
E. axelrodi Springer, 1988	2, 28	Rare, about five fish seen. Photographed.	10-40
E. bicolor (Day, 1888)	4, 6, 29, 31, 47	Occasional.	3-20
E. lividinalis Chapman and Schultz, 1952	3, 13, 17, 20, 32, 34, 51, 56	Occasional in rich coral areas, usually among branches of staghorn Acropora.	2-15
E. midas Starck, 1969	Seen on 1997 survey only.		5-30
E. namiyei (Jordan and Evermann, 1903)	8, 36, 51	Rare, about six seen.	5-30
E. pictus McKinney and Springer, 1976	14, 22	Rare, only three seen.	10-40
E. taeniatus Springer, 1988	3, 10-15, 17, 20, 21, 25, 41, 56, 57	Occasional, the most common Escenius in MBP. Photographed. Endemic to Milne Bay Province.	3-35
E. trilineatus Springer, 1972	Seen on 1997 survey only		2-20
E. yaeyamensis (Aoyagi, 1954)	10, 15, 36, 41	Rare.	1-15
Entomacrodus striatus (Quoy and Gaimard, 1836)	Seen on 1997 survey only.		0-2
Exallias brevis (Kner, 1868)	Seen on 1997 survey only.		1-20
Glyptoparus delicatulus Smith, 1959	Seen on 1997 survey only.		1-5
Laiphognathus multimaculatus Smith, 1955	Seen on 1997 survey only.		0-5
Meiacanthus anema (Bleeker, 1852)	Recorded in 1999 by GRA.		0-3
M. atrodorsalis (Günther, 1877)	1-57	Moderately common.	1-20
M. ditrema Smith-Vaniz, 1976*	37, 38, 42	Rare.	5-30
M. grammistes (Valenciennes, 1836)	1, 7-9, 11-15, 20-23, 32, 35, 38, 39, 54, 56, 57	Moderately common.	1-20
M. reticulatus Smith-Vaniz, 1976	Collected in 1972 by G. Allen.	Known from a single specimens collected in 1972 at Egum Atoll. Endemic to Milne Bay Province.	
M. vittatus Smith-Vaniz, 1976	7, 12-15, 20, 32, 34, 36, 40, 42, 46, 51, 56	Occasional.	2-15

continued

continued

Species	Site records	Abundance	Depth (m)
Parenchelyurus hepburni (Snyder, 1908)	Seen on 1997 survey only.		0-2
Petroscirtes breviceps (Valenciennes, 1836)	Recorded previously by B. Halstead.		1-10
P. variabilis Cantor, 1850	3	Rare.	1-5
Plagiotremus laudandus (Whitley, 1961)	4, 22, 24, 31, 45	Occasional.	2-35
P. rhinorhynchus (Bleeker, 1852)	1, 2, 4-7, 10-12, 14, 16, 21, 22, 25, 26, 29, 31-35, 37-39, 42-44, 46, 48-55	Common, but alway in low numbers.	1-40
P. tapeinosoma (Bleeker, 1857)	5, 37, 38, 42, 43, 47, 54	Occasional. Photographed.	1-25
Salarias fasciatus (Bloch, 1786)	10, 21, 22	Rare.	0-8
S. guttatus Valenciennes, 1836	15, 21, 36	Rare.	1-15
S. segmentatus Bath and Randall, 1991	11-13, 38, 41, 42, 57	Occasional.	2-30
Xiphasia setifer Swainson, 1839	Recorded previously by B. Halstead.		10-60
CALLIONYMIDAE			
Callionymus delicatulus Smith, 1963	Seen on 1997 survey only.		1-20
C. enneactis Bleeker, 1879	12	Collected with dipnet.	0-20
Dactylopus dactylopus (Valenciennes, 1837)	Recorded previously by B. Halstead.		1-55
Diplogrammus goramensis (Bleeker, 1858)	Recorded previously by B. Halstead.		5-35
Synchiropus moyeri Zaiser and Fricke, 1985	Seen on 1997 survey only.	Rare, 3 individuals seen.	2-20
S. splendidus (Herre, 1927)	Seen on 1997 survey only.	Collected with rotnone.	1-18
GOBIIDAE			
Amblyeleotris diagonalis Polunin and Lubbock, 1979	23	Rare, only a few seen.	6-35
A. fontanesii (Bleeker, 1852)	Seen on 1997 survey only.		5-25
A. guttata (Fowler, 1938)	1-5, 7, 12-14, 21, 23-26, 28, 31, 32, 35, 38, 39, 43, 45, 49, 50, 53, 54, 56, 57	Occasional, the most common species of shrimp goby in MBP.	10-35
A. periophthalma (Bleeker, 1853)	3, 10, 12, 15, 23, 32, 42, 48, 54, 56, 57	Occasional, locally common in some sandy areas.	8-15
A. randalli Hoese & Steene, 1978*	20, 26, 28, 33, 37, 38	Occasional on sand bottom of caves on steep slopes.	
A. sp.	Seen on 1997 survey only.		10-20
A. steinitzi (Klausewitz, 1974)	8, 10, 12, 15, 21, 23, 32-35, 39, 42, 45, 46, 48, 54, 56, 57	Occasional, locally common in some sandy areas.	6-30
A. wheeleri (Polunin and Lubbock, 1977)	1, 5, 6, 10, 31, 49, 55, 57	Occasional. Photographed.	5-20
A. yanoi Aonuma and Yoshino, 1996	5	Rare, only two individuals seen.	10-40
Ambiygobius buanensis (Herre, 1927)	Seen on 1997 survey only.		1-5

Species	Site records	Abundance	Depth (m)
A. decussatus (Bleeker, 1855)	3, 4, 7, 10-15, 17, 20-24, 26, 32, 34-36, 38, 40-43, 46, 49, 54, 56, 57	Occasional in sheltered silty areas.	3-20
A. nocturnus (Herre, 1945)	26, 46, 51	Occasional in strongly silted areas. Photographed.	3-30
A. phalaena (Valenciennes, 1837)	10, 21, 23, 29, 31-36, 38, 40, 46, 51, 56, 57	Occasional.	1-20
A. rainfordi (Whitley, 1940)	3, 4, 6, 7, 10, 13, 17, 18, 20-22, 24-28, 31-46, 48-57	Occasional, always in low numbers.	5-25
A. sphynx (Valenciennes, 1837)	Recorded previously by B. Halstead.		1-10
Asterropteryx bipunctatus Allen and Munday, 1996	11	One specimen collecetd with rotnone.	15-40
A. ensifera (Bleeker, 1874)	3	Rare.	6-40
A. semipunctatus Rüppell, 1830	23, 36	Rare, but locally common.	1-10
A. striatus Allen and Munday, 1996	1-4, 6, 14, 17, 19, 23, 38, 42, 50, 57	Occasional, but locally abundant.	5-20
Bathygobius sp.	Seen on 1997 survey only.		0-2
Bryaninops amplus Larson, 1985	4	Seen only once, but difficult to detect. No doubt common wherever seawhips are abundant.	10-40
B. loki Larson, 1985	1	Detected on only one occasion, but no doubt common where sea fans and black coral are abundant.	6-45
B. natans Larson, 1986	4, 8, 42	Rare, but relatively inconspicuous due to tiny size.	6-27
B. yongei (Davis & Cohen, 1968)	51	Seen only once, but difficult to detect. No doubt common wherever seawhips are abundant.	
Callogobius sp.	37	One specimen collected with rotenone.	3-25
Cryptocentrus cinctus (Herre, 1936)	3, 11, 46	Rare, but sand habitat not adequately surveyed.	2-15
C. fasciatus (Playfair & Günther, 1867)	Recorded previously by B. Halstead.		2-15
C. octofasciatus Regan, 1908	Seen on 1997 survey only.		1-5
C. singapurensis (Herre, 1936)	Seen on 1997 survey only.		1-5
C. strigilliceps (Jordan and Seale, 1906)	11, 21, 26, 38, 46	Rare, but sand habitat not adequately surveyed.	1-6
Ctenogobiops aurocingulus (Herre, 1935)	23	Rare, several pairs seen in silty conditions at one site.	2-15
C. feroculus Lubbock and Polunin, 1977	3, 4, 8, 10-12, 14, 21, 31, 35, 39, 43, 45, 48, 54	Occasional.	2-15
C. pomastictus Lubbock and Polunin, 1977	10, 11, 15, 24, 26, 29, 32, 34, 36, 38, 40-43, 46, 56, 57	Occasional.	2-20
C. tangaroai Lubbock and Polunin, 1977	8, 37, 43	Rare, three individuals sighted.	4-40
Discordipinna griessingeri Hoese and Fourmanoir, 1978	Seen on 1997 survey only.		5-25
Eviota afelei Jordan & Seale, 1906*	53	Two specimens collected with rotenone in 18 m depth.	1-20
E. albolineata Jewett and Lachner, 1983	3, 6, 10, 11, 35, 42	Noticed on several occasions, but easily missed due to small size.	1-10

continued

Species	Site records	Abundance	Depth (m)
E. bifasciata Lachner and Karnella, 1980	3, 4, 8, 10-15, 17, 20-23, 26-29, 32, 34, 35, 40, 51, 54, 56, 57	Occasional, but locally abundant.	5-25
E. guttata Lachner and Karanella, 1978	2, 6, 38, 42, 45, 47	Noticed on only a few occasions, but easily missed due to small size. Photographed.	3-15
E. latifasciata Jewett and Lachner, 1983	Seen on 1997 survey only.		5-20
E. melasma Lachner and Karanella, 1980	Seen on 1997 survey only.		2-15
E. nigriventris Giltay, 1933	3, 8, 10-15, 17, 20, 22, 23, 26-28, 35, 36, 41, 57	Occasional, but locally common.	4-20
E. pellucida Larson, 1976	4-18, 20-23, 26-28, 35, 37-42, 46, 51, 54, 57	Occasional.	3-20
E. prasina (Kluzinger, 1871)	Recorded previously by B. Halstead.		3-20
E. prasites Jordan and Seale, 1906	10, 13, 15, 17, 18, 25, 32, 51, 54, 56, 57	Noticed on several occasions, but easily missed due to small size.	3-15
E. punctulata Jewett and Lachner, 1983	Seen on 1997 survey only.		1-10
E. queenslandica Whitley, 1932	10, 42	Noticed on only two occasions, but easily missed due to small size.	5-30
E. sebreei Jordan and Seale, 1906	4, 6, 27, 35, 38	Noticed on only a few occasions, but easily missed due to small size.	3-20
E. sparsa Jewett and Lachner, 1983	Seen on 1997 survey only.		-
Exyrias bellisimus (Smith, 1959)	13-15, 22, 26, 38, 51, 56	Occasional on silty reefs. Photographed.	1-25
Fusigobius duospilus Hoese and Reader, 1985	29, 37, 53		-
F. longispinus Goren, 1978	Seen on 1997 survey only.		8-25
F. neophytus (Günther, 1877)	15, 21, 27, 32, 35, 46, 48, 49, 54	Occasional.	2-15
F. signipinnis Hoese and Obika, 1988	4, 10-14, 20-22, 24, 25, 35, 37-46, 49, 50, 54, 57	Occasional.	10-30
F. sp. 1	26	Rare. Translucent with large internal dark spot on caudal peduncle.	5-25
F. sp. 2	25	One collected with rotenone. Translucent with orange spots.	5-25
Gobiopsis sp.	Recorded previously by B. Halstead.		5-15
Gnatholepis cauerensis (Bleeker, 1853)	11, 38	Only a few seen, but easily escapes notice due to small size and cryptic habits. Photographed.	1-45
G. scapulostigma Herre, 1953	4, 20, 22, 29, 31, 35, 39, 41, 43, 45, 46, 48, 54, 57	Occasional.	3-30
Gobiodon citrinus (Rüppell, 1838)*	8	Only a few seen, but easily escapes notice due to small size and cryptic habits.	1-15
G. histrio (Valenciennes, 1837)	Seen on 1997 survey only.		3-12

continued

Species	Site records	Abundance	Depth (m)
G. okinawae Sawada, Arai and Abe, 1973	10, 11, 21, 26, 38	Relatively rare, but a secretive species that is easily overlooked.	2-12
G. quinquestrigattus (Valenciennes, 1837)*	46	Only a few seen, but easily escapes notice due to small size and cryptic habits.	2-12
G. spilophthalmus Fowler, 1944	Seen on 1997 survey only.	Previously reported as G. albofasciatus.	2-15
Istigobius decoratus (Herre, 1927)	15, 21, 24, 26, 27, 34, 40, 41, 57	Occasional.	1-18
I. ornatus (Rüppell, 1830)	Seen on 1997 survey only.		0-5
I. rigilius (Herre, 1953)	4, 10, 11, 23, 31, 35, 39, 43-45, 48-50, 53, 54, 56	Occasional.	0-30
Lotila graciliosa Klausewitz, 1960	4	Rare.	2-15
Lubricogobius sp.	Recorded previously by B. Halstead.		2-18
Luposicya lupus Smith, 1959	Recorded previously by B. Halstead.		2-18
Macrodontogobius wilburi Herre, 1936	11, 15, 26, 38, 46, 51, 56	Occasional in silty areas. Photographed.	2-15
Oplopomops atherinoides (Peters, 1855)*	46	Rare, only one adult seen.	2-15
Oplopomus diacanthus (Schultz, 1943)*	23	Probably common, but seldom noticed in sandy areas. Photographed.	1-10
O. oplopomus (Valenciennes, 1837)	Seen on 1997 survey only.		2-25
Oxyurichthys ophthalmolepis Bleeker, 1856	Seen on 1997 survey only.		10-50
O. papuensis (Valenciennes, 1839)	Seen on 1997 survey only.		1-8
Paragobiodon echinocephalus (Rüppell, 1830)*	6	Only a few seen, but easily escapes notice due to small size and cryptic habits.	1-12
P. melanosomus (Bleeker, 1852)	Seen on 1997 survey only.	.	1-12
P. xanthosomus (Bleeker, 1852)	Seen on 1997 survey only.		1-10
Phyllogobius platycephalops (Smith, 1964)*	46	Only a few seen, but easily escapes notice due to small size and cryptic habits. Commensal with sponges (Phyllospongia).	3-20
Pleurosicya annadalei Hornell and Fowler, 1922	Recorded previously by B. Halstead.		30-65
P. elongata Larson, 1990*	3, 13, 15, 21, 26, 28, 29, 34, 35, 38, 42, 43, 46, 56, 57	Occasional, commensal with sponge (Ianthella basta.).	10-40
P. micheli Fourmanoir, 1971	Recorded previously by B. Halstead.		10-50
P. mossambica Smith, 1959	Recorded previously by B. Halstead.		1-35
P. spongicola Larson, 1990	Recorded previously by B. Halstead.		10-40
Priolepis cincta (Regan, 1908)	Seen on 1997 survey only.		1-70
P. fallacincta Winterbottom & Burridge, 1992*	45	One specimen collected with rotenone.	

continued

continued

Species	Site records	Abundance	Depth (m)
P. semidoliatus (Valenciennes, 1837)	Seen on 1997 survey only.		0-10
Signigobius biocellatus Hoese and Allen, 1977	7, 12-14, 26, 36, 38, 40, 42	Occasional on silty bottoms.	2-30
Stonogobiops xanthorhinica Hoese and Randall, 1982	Recorded previously by B. Halstead.		12-60
Tomiyamichthys oni (Tomiyama, 1936)*	42	Rare, only one seen.	5-30
Trimma benjamini Winterbottom, 1996	37, 53	Several specimens collected with rotenone. Photographed.	10-24
T. caesiura (Jordan and Seale, 1906)	20, 38, 40, 42, 53	Rare, but easily overlooked due to small size and secretive habits.	2-12
T. emeryi Winterbottom, 1984*	22, 25	Four specimens collected with rotenone.	
T. griffithsi Winterbottom, 1984	8, 12-14, 16, 17, 38, 41, 42	Occasional, but is easily overlooked due to small size and secretive habits.	20-40
T. hoesei Winterbottom, 1984	22, 37	Two specimens collected with rotenone. Photographed.	15-30
T. macrophthalma (Tomiyama, 1936)	53	Three specimens collected with rotenone.	5-30
T. naudei Smith, 1957*	53	Two specimens collected with rotenone.	
T. okinawae (Aoyagi, 1949)	53	Nine specimens collected with rotenone.	5-30
T. rubromaculata Allen and Munday, 1995	10, 25	Rare, only four individuals seen.	20-35
T. sp. 1 (dusky reddish with small black "ear" spot)	13, 35, 37, 53	Collected with rotenone.	8-40
T. sp. 2 (overall yellow with large black spot of first dorsal fin)	Seen on 1997 survey only.		8-25
T. sp. 3 (yellow to reddish with black peduncle and with tail)	53	Three specimens collected with rotenone.	5-20
T. sp. 4 (broad alternating red and white bars)	37	Rare, several seen at one site.	10-30
T. sp. 5 (pinkish, Gobiodon-like shape)	53	One collected with rotenone.	5-15
T. sp. 6 (reddish with double row of large orange blotches)*	53	Two specimens collected with rotenone.	5-25
T. sp.7 (reddish with 3-4 rows of irregular pale blotches)*	53	One specimen collected with rotenone.	5-25
T. sp. 8 (red with yellow mid-lateral stripe, white on belly)*	10	One specimen collected with quinaldine sulphate.	25-40
T. striata (Herre, 1945)	4	Rare, but easily overlooked due to small size and secretive habits.	2-25
T. taylori Lobel, 1979	53	Three specimens collected with rotenone.	15-50
T. tevegae Cohen and Davis, 1969	2, 5, 6, 8-10, 12-14, 16-20, 22, 24-28, 35, 37, 38, 40, 41, 43, 44, 49, 52, 53, 57	In spite of its small size and cryptic habits it appears to be common under ledges and in caverns on vertical slopes.	8-45

Species	Site records	Abundance	Depth (m)
Trimmatom nanus Winterbottom and Emery, 1981	Seen on 1997 survey only.		6-35
T. sp.	Seen on 1997 survey only.		5-25
Valenciennea helsdingenii (Bleeker, 1858)	Seen on 1997 survey only.		1-30
V. longipinnis (Lay & Bennett, 1839)*	39	Rare, one large adult seen.	1-20
V. muralis (Valenciennes, 1837)	15, 21, 23, 36, 41, 42, 46, 56, 57	Occasional in shallow sandy areas.	1-15
V. puellaris (Tomiyama, 1936)	1, 29, 31, 32, 34, 35, 39, 43	Occasional. Photographed.	2-30
V. randalli Hoese and Larson, 1994	42	Rare, several seen in 25 m depth.	8-30
V. sexguttata (Valenciennes, 1837)	10, 11	Rare, about 10 seen.	1-10
V. strigata (Broussonet, 1782)	4, 14, 18, 24, 26, 29, 31, 32, 35, 37, 39, 44, 45, 47, 50	Occasional, in relatively low numbers at each site.	1-25
V. wardii (Playfair, 1867)	Recorded previously by B. Halstead.		5-30
Vanderhorstia ambanoro (Fourmanoir, 1957)	Seen on 1997 survey only.		4-20
V. flavolineata Allen & Munday, 1995*	25	Rare, two seen in 35 m depth.	25-50
V. ornatissima Smith, 1959	3	Rare.	15-20
MICRODESMIDAE			
Aioliops megastigma Rennis and Hoese, 1987	6-12, 15, 17, 20-22, 32, 34-36, 40, 41, 46, 51, 56	Occasional.	1-15
Gunnelichthys curiosus Dawson, 1968	Seen on 1997 survey only.		2-30
G. monostigma Smith, 1958	Recorded previously by B. Halstead.		2-20
G. viridescens Dawson, 1968	Recorded previously by B. Halstead.		2-20
Nemateleotris decora Randall and Allen, 1973	Seen on 1997 survey only.		28-70
N. magnifica Fowler, 1938	2, 5, 6, 12, 30, 37, 39, 45, 47, 49, 50, 53	Occasional.	6-61
Oxymetopon cyanoctenosum Klausewitz and Condé, 1981	Seen on 1997 survey only.		20-30
Parioglossus formosus (Smith, 1931)*	24	Locally common, but easily overlooked. Three collected.	
P. nudus Rennis and Hoese, 1985	24-26, 38	Rare, but easily overlooked due to small size. Photographed.	10-35
P. palustris (Herre, 1945)	Seen on 1997 survey only.		0-2
Ptereleotris evides (Jordan and Hubbs, 1925)	3, 7, 10, 14, 15, 18, 22, 27, 29-31, 33, 36, 38, 42-45, 47, 49-51, 54, 55	Moderately common.	2-15
P. grammica Randall and Lubbock, 1982	Recorded previously by B. Halstead.		35-50
P. hanae (Jordan and Synder, 1901)	Recorded previously by B. Halstead.		3-43
P. heteroptera (Bleeker, 1855)	6, 9, 16, 19, 47	Occasional, usually below 25 m depth.	6-50
P. microlepis Bleeker, 1856	36, 38, 39, 46, 51, 56, 57	Occasional.	1-22

continued

Species	Site records	Abundance	Depth (m)
P. monoptera Randall and Hoese, 1985	34	Rare, only a single pair seen in 15 m depth.	5-20
P. uroditaenia Randall and Hoese, 1985	Recorded previously by B. Halstead.		10-30
P. zebra (Fowler, 1938)	44, 45	Rare.	2-10
XENISTHMIDAE			
Xenisthmus polyzonatus (Klunzinger, 1871)	Seen on 1997 survey only.		5-20
EPHIPPIDAE			
Platax batavianus (Cuvier, 1831)	56	Rare, one large adult seen. Photographed.	1-40
P. boersi Bleeker, 1852	3, 9, 12, 14, 15, 18, 22, 26, 37, 38, 40, 47	Occasional.	1-20
P. orbicularis (Forsskål, 1775)	7, 21, 29, 37	Occasional.	1-30
P. pinnatus (Linnaeus, 1758)	6, 7, 10, 12, 14, 20, 22, 30, 32, 35, 39, 41, 42, 51, 56	The most common batfish encountered, but only occasional sightings. Photographed.	1-35
P. teira (Forsskål, 1775)	4, 30, 37, 49, 57	Rarely seen.	0-2
SCATOPHAGIDAE			
Scatophagus argus (Bloch, 1788)	Seen on 1997 survey only.		0-10
SIGANIDAE			
Siganus argenteus (Quoy and Gaimard, 1824)	2, 7, 11, 12, 19, 21, 27, 29-31, 41, 42, 47-49, 56	Occasional, but locally common.	1-30
S. corallinus (Valenciennes, 1835)	1, 3-9, 12-22, 24-33, 35-39, 45-50, 54-57	Moderately common.	4-25
S. doliatus Cuvier, 1830	3, 11, 12, 14-16, 18, 22, 26, 32-36, 38, 40-42, 46, 56, 57	Occasional, but locally common.	1-15
S. fuscescens (Houttyn, 1782)	3	Rare, but locally common where seagrass abundant.	0-15
S. lineatus (Linnaeus, 1835)	3, 14, 20, 22, 23, 34, 36	Occasional.	1-25
S. puellus (Schlegel, 1852)	2, 4, 6-14, 16-22, 26, 29, 30, 32, 33, 35-39, 41-43, 47-51, 54, 55, 57	Common.	2-30
S. punctatissimus Fowler and Bean, 1929	1, 4-8, 11, 12, 14, 16, 17, 19, 21, 22, 25, 28, 32, 34, 35, 40-43, 45, 46, 55-57	Common.	3-30
S. punctatus (Forster, 1801)	4, 6, 9, 10, 12, 19, 30, 31, 35, 37, 40, 44, 45, 49, 50, 56	Moderately common.	1-40
S. randalli Woodland, 1990	24	Rare, a group of five seen in 12 m depth.	1-15
S. spinus (Linnaeus, 1758)	8, 10, 15, 56	Occasional.	1-12
S. vulpinus (Schlegel and Müller, 1844)	1-7, 9-33, 35-44, 47-51, 54, 56, 57	Common.	1-30
ZANCLIDAE			
Zanclus cornutus Linnaeus, 1758	1-56	Common.	1-180

continued

Species	Site records	Abundance	Depth (m)
ACANTHURIDAE			
Acanthurus bariene Lesson, 1830*	44	Rare, several seen on outer reef in 20 m depth. Photographed.	15-50
A. blochi Valenciennes, 1835	1-3, 7, 10, 19, 21, 24, 27, 30, 31, 35, 36, 39, 40, 43, 45, 48-51, 54	Common.	3-20
A. dussumieri Valenciennes, 1835	29-31, 37, 39, 40, 43, 45, 47, 49, 51, 53	Occasional. Seen only at Conflict Group and Louisiades. Photographed.	10-30
A. fowleri de Beaufort, 1951	1, 2, 4, 6, 16-20, 24-28, 57	Occasional, but absent from Louisiades.	10-30
A. leucocheilus Herre, 1927	Seen on 1997 survey only.		5-20
A. lineatus (Linnaeus, 1758)	4, 6, 8, 11, 12, 14, 15, 18, 20-22, 24-31, 33, 36, 39-45, 47-53, 55, 56	Moderately common, usually in shallow surge-affected areas.	1-15
A. maculiceps (Ahl, 1923)*	52	Rare, only one adult seen.	1-15
A. mata (Cuvier, 1829)	2, 4-7, 9, 24, 25, 27, 28, 30, 33, 50, 52	Moderately common, usually on dropoffs in turbid water.	5-30
A. nigricans (Linnaeus, 1758)	4, 5, 30, 43-45, 49, 50, 52, 53	Generally rare, but locally common, particularly at Louisiades.	3-65
A. nigricaudus Duncker and Mohr, 1929	1, 3, 8, 10-17, 19, 21-23, 29-39, 41-50, 52, 54, 56, 57	Moderately common.	3-30
A. nigrofuscus (Forsskål, 1775)	12, 29-32, 37-39, 41, 44, 45, 51, 52, 54	Occasional.	2-20
A. nubilus (Fowler and Bean, 1929)	2, 24, 25	Rare, about five seen.	10-30
A. olivaceus Bloch and Schneider, 1801	6, 10, 19, 45, 47, 54	Occasional, but common at site 47, where spawning observed.	5-45
A. pyroferus Kittlitz, 1834	1-5, 7-10, 12-22, 24-35, 37-39, 42-45, 57-50, 52-57	Common.	4-60
A. thompsoni (Fowler, 1923)	2, 5, 6, 9, 10, 24, 25, 29, 30, 37, 39, 43, 44, 49, 50, 52-54	Moderately common, usually on steep dropoffs.	4-75
A. triostegus (Linnaeus, 1758)	4, 7, 8, 10, 12, 31, 32, 37, 51-53	Occasional, usually in shallow wave-affected areas. More common in Louisiades.	0-90
A. xanthopterus Valenciennes, 1835	1, 2, 4, 7, 9, 11, 21-23, 33, 35, 36, 38, 43, 46, 51, 53, 54, 56	Occasional, usually on sandy slopes adjacent to reefs.	3-90
Ctenochaetus binotatus Randall, 1955	1-24, 29-36, 38, 40-57	Common.	10-55
C. striatus (Quoy and Gaimard, 1824)	1, 3-7, 9-57	Common, usually in depths less than 10 m.	2-30
C. strigosus (Bennett, 1828)	2, 4, 9, 14, 17, 29, 37, 39, 43	Only one noticed, but hard to differentiate from *C. striatus* at a distance.	3-25
C. tominiensis Randall, 1955	1-4, 6, 7, 11-15, 18, 20-28, 35, 37, 38, 42, 46, 57	Moderately common, especially in sheltered locations that drop steeply to deep water.	5-40

continued

Species	Site records	Abundance	Depth (m)
Naso annulatus (Quoy and Gaimard, 1825)	2, 4, 9, 33, 37, 39, 47, 52	Moderately common, usually adjacent to steep outer slopes.	15-40
N. brachycentron (Valenciennes, 1835)	4, 5, 29, 46, 47, 55	Occasional.	15-50
N. brevirostris (Valenciennes, 1835)	2, 5, 6, 8-10, 16-18, 26, 29, 30, 37, 39, 44, 46-50, 54	Moderately common.	4-50
N. hexacanthus (Bleeker, 1855)	2, 5, 6, 8, 9, 16, 19, 28, 37, 39, 52-54	Occasional, but locally common to abundant.	6-140
N. lituratus (Bloch and Schneider, 1801)	1-22, 24-57	Common.	5-90
N. lopezi Herre, 1927	2, 19, 24	Rare.	6-70
N. minor (Smith, 1966)	Seen on 1997 survey only.		10-50
N. thynnoides (Valenciennes, 1835)	2, 19, 26, 50	Generally rare, but several large schools seen.	8-50
N. tuberosus Lacepède, 1801	29, 30, 39, 40, 47, 50, 51	Occasional.	3-20
N. unicornis (Forsskål, 1775)	4, 8, 11, 13, 14, 16, 21, 28, 30, 31, 35, 40, 41, 43, 45, 46, 49, 51, 55	Moderately common.	4-80
N. vlamingii Valenciennes, 1835	2, 4-6, 9, 11, 12, 15-19, 23, 25, 27-30, 34, 35, 37-39, 44-47, 49, 50, 52-54	Modertaely common, adjacent to steeper outer slopes.	4-50
Paracanthurus hepatus (Linnaeus, 1758)	Seen on 1997 survey only.		2-40
Zebrasoma scopas (Cuvier, 1829)	1-57	Common:	1-60
Z. veliferum (Bloch, 1797)	1-57	Common.	4-30
SPHYRAENIDAE			
Sphyraena barracuda (Walbaum, 1792)	16, 30, 45, 56	Rare, about five seen. Also caught by trolling from charter vessel.	0-20
S. flavicauda Rüppell, 1838	38, 42, 56	Occasional in schools of 5-50 fish.	1-20
S. jello Cuvier, 1829	4, 6, 16	Occasional in schools of upt to 50 individuals.	1-20
S. qenie Klunzinger, 1870	16	Rare.	5-40
SCOMBRIDAE			
Euthynnus affinis (Cantor, 1849)	2, 54	Rarely sighted.	0-20
Grammatorcynus bicarinatus (Quoy & Gaimard, 1824)*	43	Rare, only one seen.	5-40
G. bilineatus (Quoy and Gaimard, 1824)	2, 6, 9, 19, 23, 31, 32, 44, 52	Occasional.	10-40
Gymnosarda unicolor (Rüppell, 1836)	6, 8, 9, 12, 13, 19, 34-36, 49, 52	Occasional.	5-100
Rastrelliger kanagurta (Cuvier, 1816)	14, 22, 47	Occasional, but usually in large schools.	0-30
Scomberomorus commerson (Lacepède, 1800)	6, 37	Occasional individuals seen.	0-30

continued

Species	Site records	Abundance	Depth (m)
BOTHIDAE			
Bothus maricus (Broussonet, 1782)	31	Only one seen, but very difficult to detect due to camouflage coloration.	5-30
Sterhombus intermediua (Bleeker, 1866)	Recorded previously by B. Halstead.		5-15
PLEURONECTIDAE			
Samariscus triocellattus Woods, 1960	Recorded previously by B. Halstead.		10-50
BALISTIDAE			
Abalistes stellatus (Lacepède, 1798)	14, 42	Rare, only one juvenile seen.	10-120
Balistapus undulatus (Park, 1797)	1-57	Common.	3-50
Balistoides conspicillum (Bloch and Schneider, 1801)	2, 13, 14, 16, 19, 25, 29, 30, 44, 45, 47, 49, 50, 52-55	Occasional.	10-50
B. viridescens (Bloch and Schneider, 1801)	1, 3, 6, 10, 12, 14, 16, 18, 19, 25, 26, 29-31, 36, 39, 41, 43-50, 54	Occasional.	5-45
Canthidermis maculatus (Bloch, 1786)	2, 6	Rarely seen, but locally common on steeper outer slopes.	1-30
Melichthys vidua (Solander, 1844)	2, 4-6, 9, 14, 16, 19, 23, 29, 30, 39, 47, 49, 50, 52, 53	Occasional.	3-60
Odonus niger (Rüppell, 1836)	2, 4-6, 9, 16, 19, 23, 25-29, 33, 39	Occasional, but locally common at some sites.	3-40
Pseudobalistes flavimarginatus (Rüppell, 1828)	6, 7, 36, 46, 48	Occasional, in sheltered sand or rubble areas.	2-50
Rhinecanthus aculeatus (Linnaeus, 1758)	51	Rare, about five seen.	0-3
R. rectangulus (Bloch and Schneider, 1801)	4, 52, 53	Rare, less than 10 encountered.	1-3
R. verrucosus (Linnaeus, 1758)	7, 10-12, 15, 42, 47	Occasional, but locally common on shallow flats near shore.	0-3
Sufflamen bursa (Bloch and Schneider, 1801)	1, 2, 4, 7, 8, 14, 15, 17, 19, 20, 22-40, 43-45, 48-50, 52-55	Occasional.	3-90
S. chrysoptera (Bloch and Schneider, 1801)	3-7, 12, 19, 29, 31-33, 35, 38-40, 42-45, 47, 51, 54, 55	Occasional.	1-35
S. fraenatus (Latreille, 1804)	Recorded previously by B. Halstead.		8-185
MONACANTHIDAE			
Acreichthys tomentosus (Linnaeus, 1758)*	14	Rare, but weed habitat not surveyed.	1-10
Aluterus scriptus (Osbeck, 1765)	4, 6, 43	Circumtropical. Rare, only three observed.	2-80
Amanses scopas (Cuvier, 1829)	6, 8, 9, 12, 17, 19, 27, 54	Rare, about 10 seen.	3-20
Cantherines dumerilii (Hollard, 1854)	Seen on 1997 survey only.		1-35
C. fronticinctus (Günther, 1866)	1, 10, 17, 19, 50, 54	Rare.	2-40
C. pardalis (Rüppell, 1866)	Recorded previously by B. Halstead.		2-20
Oxymonacanthus longirostris (Bloch and Schneider, 1801)	4, 6, 7-9, 11, 18, 29	Occasional, in rich areas.	1-30

continued

Species	Site records	Abundance	Depth (m)
Paraluteres prionurus (Bleeker, 1851)	23, 30, 35	Rare, only three fish seen.	2-25
Pervagor janthinosoma (Bleeker, 1854)	8, 9	Rare, only two seen.	2-18
P. melanocephalus (Bleeker, 1853)	17, 30, 49	Rare, only three seen.	15-40
P. nigrolineatus (Herre, 1927)	Seen on 1997 survey only.	Rare, only 3 seen.	2-15
Pseudaluteres nasicornis (Schlegel, 1846)	Recorded previously by B. Halstead.		1-55
Pseudomonacanthus macrurus (Bleeker, 1856)	Recorded previously by B. Halstead.		5-40
Rudarius excelsus Hutchins, 1977	Recorded previously by B. Halstead.		2-15
R. minutus Tyler, 1970	Recorded previously by B. Halstead.	Rare, one group containing 4 individuals seen.	2-15
OSTRACIIDAE			
Lactoria cornuta (Linnaeus, 1758)	Seen on 1997 survey only.		1-50
Ostracion cubicus Linnaeus, 1758	4, 12, 15, 30, 31, 36, 38, 43, 50	Occasional.	1-40
O. meleagris Shaw, 1796	2, 9, 29, 30, 33, 39, 43, 44, 49, 50, 54	Occasional.	2-30
O. solorensis Bleeker, 1853	2, 24-26, 30, 44, 50	Occasional.	1-20
TETRAODONTIDAE			
Arothron caeruleopunctatus Matsuura, 1994	Seen on 1997 survey only.		5-30
A. hispidus (Linnaeus, 1758)	35	Rare, only one seen.	1-50
A. manilensis (Marion de Procé, 1822)	Seen on 1997 survey only.		1-20
A. mappa (Lesson, 1830)	23, 31, 35, 39, 43	Rare, five individuals seen.	4-40
A. nigropunctatus (Bloch and Schneider, 1801)	1, 2, 4, 6, 7, 10, 11, 13-15, 17-19, 26, 28-30, 35-37, 43, 44, 49, 50, 52, 54-57	Moderately common, but always in low numbers at each site.	2-35
A. stellatus (Schneider, 1801)	26, 38	Rare, two large adult seen.	3-58
Canthigaster bennetti (Bleeker, 1854)	7, 39, 45	Rare, only three pairs seen.	1-10
C. compressa (Marion de Procé, 1822)	Seen on 1997 survey only.		1-20
C. epilampra (Jenkins, 1903)	Recorded previously by B. Halstead.		3-20
C. janthinoptera (Bleeker, 1855)	Seen on 1997 survey only		9-60
C. ocellicincta Allen and Randall, 1977	Recorded previously by B. Halstead.		10-30
C. solandri (Richardson, 1844)	2, 8, 9, 12, 14, 19, 23, 24, 26, 27, 36, 41, 51, 56, 57	Occasional.	1-36
C. valentini (Bleeker, 1853)	4-6, 9, 23, 29-31, 35, 37-39, 42-44, 48-51, 53-57	Occasional.	3-55
DIODONTIDAE			
Diodon hystrix Linnaeus, 1758	14, 18, 31	Rare, only three seen.	1-30
D. liturosus Shaw, 1804	Seen on 1997 survey only.		3-35

Top Fish Diversity Sites in MBP (with species numbers)

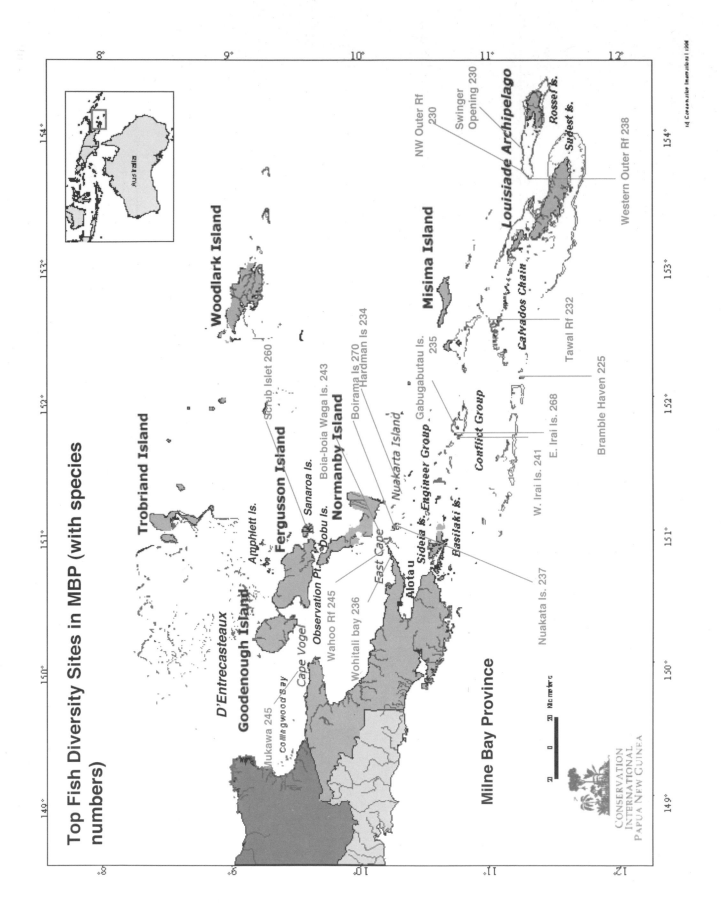

Milne Bay Province

CONSERVATION
INTERNATIONAL
PAPUA NEW GUINEA

Appendix 6

List of reef fish target species of Milne Bay Province, Papua New Guinea

Mark Allen

Summary of Beche-de-mer harvesting areas are presented in the map following the table.

No.	Family/species	No. sites where present	Percent occurrence
I	**HEMISCYLIIDAE**		
1	*Hemiscyllium trispeculare* Richardson, 1843	1	2%
II	**GINGLYMOSTOMATIDAE**		
2	*Nebrius ferrugineus* (Lesson, 1830)	1	2%
III	**CARCHARHINIDAE**		
3	*Carcharhinus albimarginatus* (Rüppell, 1837)	1	2%
4	*C. amblyrhynchos* (Bleeker, 1856)	4	8%
5	*Triaenodon obesus* (Rüppell, 1835)	4	8%
IV	**DASYATIDIDAE**		
6	*Dasyatis kuhlii* (Müller and Henle, 1841)	1	2%
7	*Taeniura lymma* (Forsskål, 1775)	4	8%
8	*Taeniura meyeni* (Müller and Henle, 1841)	1	2%
V	**MYLIOBATIDAE**		
9	*Aetobatus narinari* (Euphrasen, 1790)	1	2%
VI	**CLUPEIDAE**		
10	*Amblygaster sirm* (Walbaum, 1792)	1	2%
VII	**PLOTOSIDAE**		
11	*Plotosus lineatus* (Thünberg, 1787)	5	10%
VIII	**HOLOCENTRIDAE**		
12	*Myripristis adusta* Bleeker, 1853	9	18%
13	*M. berndti* Jordan and Evermann, 1902	11	22%
14	*M. hexagona* (Lacepède, 1802)	2	4%
15	*M. kuntee* Valenciennes, 1831	18	36%
16	*M. murdjan* (Forsskål, 1775)	6	12%
17	*M. violacea* Bleeker, 1851	29	58%
18	*M. vittata* Valenciennes, 1831	3	6%
19	*Neoniphon opercularis* (Valenciennes, 1831)	2	4%
20	*N. sammara* (Forsskål, 1775)	19	38%
21	*Sargocentron caudimaculatum* (Rüppell, 1835)	7	14%
22	*S. cornutum* (Bleeker, 1853)	1	2%
23	*S. diadema* (Lacepède, 1802)	5	10%
24	*S. melanospilos* (Bleeker, 1858)	1	2%
25	*S. rubrum* (Forsskål, 1775)	3	6%
26	*S. spiniferum* (Forsskål, 1775)	13	26%
27	*S. violaceus* (Bleeker, 1853)	2	4%
IX	**PLATYCEPHALIDAE**		
28	*Cymbacephalus beauforti* Knapp, 1973	2	4%
X	**SERRANIDAE**		
29	*Aethaloperca rogaa* (Forsskål, 1775)	7	14%
30	*Anyperodon leucogrammicus* (Valenciennes, 1828)	16	32%
31	*Cephalopholis argus* Bloch and Schneider, 1801	9	18%
32	*C. boenack* (Bloch, 1790)	5	10%
33	*C. cyanostigma* (Kuhl and Van Hasselt, 1828)	27	54%
34	*C. leopardus* (Lacepède, 1802)	6	12%
35	*C. microprion* (Bleeker, 1852)	23	46%
36	*C. miniata* (Forsskål, 1775)	13	26%
37	*C. sexmaculata* Rüppell, 1828	7	14%
38	*C. urodeta* (Schneider, 1801)	11	22%

continued

No.	Family/species	No. sites where present	Percent occurrence
39	*Cromileptes altivelis* (Valenciennes, 1828)	8	16%
40	*Epinephelus caruleopunctatus* (Bloch, 1790)	1	2%
41	*E. corallicola* (Kuhl and Van Hasselt, 1828)	2	4%
42	*E. cyanopodus* (Richardson, 1846)	3	6%
43	*E. fasciatus* (Forsskål, 1775)	10	20%
44	*E. fuscoguttatus* (Forsskal, 1775)	2	4%
45	*E. maculatus* (Bloch, 1790)	5	10%
46	*E. merra* Bloch, 1793	10	20%
47	*E. ongus* (Bloch, 1790)	7	14%
48	*E. polyphekadion* (Bleeker, 1849)	2	4%
49	*Gracila albimarginata* (Fowler and Bean, 1930)	4	8%
50	*Plectropomus areolatus* (Rüppell, 1830)	2	4%
51	*P. laevis* (Lacepède, 1802)	13	26%
52	*P. leopardus* (Lacepède, 1802)	19	38%
53	*P. oligocanthus* (Bleeker, 1854)	21	42%
54	*Variola albimarginata* Baissac, 1953	8	16%
55	*V. louti* (Forsskål, 1775)	6	12%
XI	**PRIACANTHIDAE**		
56	*Priacanthus hamrur* (Forsskål, 1775)	2	4%
XII	**ECHENEIDAE**		
57	*Echeneis naucrates* Linnaeus, 1758	2	4%
XIII	**CARANGIDAE**		
58	*Carangoides bajad* (Forsskål, 1775)	22	44%
59	*C. ferdau* (Forsskål, 1775)	5	10%
60	*C. fulvoguttatus* (Forsskål, 1775)	2	4%
61	*C. plagiotaenia* Bleeker, 1857	14	28%
62	*Caranx ignobilis* (Forsskål, 1775)	1	2%
63	*C. melampygus* Cuvier, 1833	15	30%
64	*C. sexfasciatus* Quoy and Gaimard, 1825	2	4%
65	*Elegatis bipinnulatus* (Quoy and Gaimard, 1825)	4	8%
66	*Scomberoides lysan* (Forsskål, 1775)	1	2%
XIV	**LUTJANIDAE**		
67	*Aphareus furca* (Lacepède, 1802)	4	8%
68	*Aprion virescens* Valenciennes, 1830	3	6%

No.	Family/species	No. sites where present	Percent occurrence
69	*Lutjanus biguttatus* (Valenciennes, 1830)	22	44%
70	*L. bohar* (Forsskål, 1775)	34	68%
71	*L. boutton* (Lacepède, 1802)	1	2%
72	*L. carponotatus* (Richardson, 1842)	10	20%
73	*L. fulviflamma* (Forsskål, 1775)	1	2%
74	*L. fulvus* (Schneider, 1801)	9	18%
75	*L. gibbus* (Forsskål, 1775)	20	40%
76	*L. kasmira* (Forsskål, 1775)	4	8%
77	*L. monostigma* (Cuvier, 1828)	10	20%
78	*L. quinquelineatus* (Bloch, 1790)	8	16%
79	*L. rivulatus* (Cuvier, 1828)	3	6%
80	*L. russelli* (Bleeker, 1849)	4	8%
81	*L. semicinctus* Quoy and Gaimard, 1824	26	52%
82	*L. vitta* (Quoy and Gaimard, 1824)	2	4%
83	*Macolor macularis* Fowler, 1931	35	70%
84	*M. niger* (Forsskål, 1775)	3	6%
85	*Paracaesio sordidus* Abe & Shinohara, 1962	1	2%
86	*Symphorichthys spilurus* (Günther, 1874)	3	6%
87	*Symphorus nematophorus* (Bleeker, 1860)	2	4%
XV	**CAESIONIDAE**		
88	*Caesio caerulaurea* Lacepède, 1802	22	44%
89	*C. cuning* (Bloch, 1791)	31	62%
90	*C. lunaris* Cuvier, 1830	12	24%
91	*C. teres* Seale, 1906	6	12%
92	*Gymnocaesio gymnoptera* (Bleeker, 1856)	5	10%
93	*Pterocaesio digramma* (Bleeker, 1865)	19	38%
94	*P. pisang* (Bleeker, 1853)	31	62%
95	*P. tessellata* Carpenter, 1987	4	8%
96	*P. tile* (Cuvier, 1830)	6	12%
97	*P. trilineata* Carpenter, 1987	9	18%
XVI	**NEMIPTERIDAE**		
98	*Pentapodus* sp.	15	30%
99	*P. trivittatus* (Bloch, 1791)	4	8%
100	*Scolopsis bilineatus* (Bloch, 1793)	35	70%
101	*S. ciliatus* (Lacepède, 1802)	1	2%

continued

No.	Family/species	No. sites where present	Percent occurrence
102	*S. lineatus* Quoy and Gaimard, 1824	2	4%
103	*S. margaritifer* (Cuvier, 1830)	28	56%
104	*S. monogramma* (Kuhl and Van Hasselt, 1830)	1	2%
105	*S. temporalis* (Cuvier, 1830)	2	4%
XVII	**HAEMULIDAE**		
106	*Plectorhinchus celebicus* Bleeker, 1873	5	10%
107	*P. chaetodontoides* (Lacepède, 1800)	20	40%
108	*P. gibbosus* (Lacepède, 1802)	2	4%
109	*P. lessoni* (Cuvier, 1830)	1	2%
110	*P. lineatus* (Linnaeus, 1758)	4	8%
111	*P. obscurus* (Günther, 1871)	2	4%
112	*P. orientalis* (Bloch, 1793)	2	4%
XVIII	**LETHRINIDAE**		
113	*Gnathodentex aurolineatus* Lacepède, 1802	1	2%
114	*G.* sp.	4	8%
115	*Lethrinus erythracanthus* Valenciennes, 1830	17	34%
116	*L. erythropterus* Valenciennes, 1830	14	28%
117	*L. obsoletus* (Forsskål, 1775)	4	8%
118	*L. olivaceous* Valenciennes, 1830	4	8%
119	*L. xanthocheilus* Klunzinger, 1870	1	2%
120	*Monotaxis grandoculis* (Forsskål, 1775)	48	96%
XIX	**MULLIDAE**		
121	*Mulloidichthys flavolineatus* (Lacepède, 1802)	1	2%
122	*M. vanicolensis* (Valenciennes, 1831)	1	2%
123	*Parupeneus barberinoides* (Lacepède, 1801)	1	2%
124	*P. barberinus* (Lacepède, 1801)	39	78%
125	*P. bifasciatus* (Lacepède, 1801)	32	64%
126	*P. cyclostomus* (Lacepède, 1802)	12	24%
127	*P. heptacanthus* (Lacepède, 1801)	1	2%
128	*P. indicus* (Shaw, 1903)	1	2%
129	*P. multifasciatus* Bleeker, 1873	34	68%
130	*P. pleurostigma* (Bennett, 1830)	4	8%
131	*Upeneus tragula* Richardson, 1846	5	10%
XX	**KYPHOSIDAE**		
132	*Kyphosus cinerascens* (Forsskål, 1775)	2	4%
133	*K. vaigiensis* (Quoy and Gaimard, 1825)	3	6%
XXI	**EPHIPPIDAE**		
134	*Platax boersi* Bleeker, 1852	9	18%
135	*P. pinnatus* (Linnaeus, 1758)	8	16%
136	*P. teira* (Forsskål, 1775)	5	10%
XXII	**SPHYRAENIDAE**		
137	*Sphyraena barracuda* (Walbaum, 1792)	1	2%
138	*S. flavicauda* Rüppell, 1838	1	2%
139	*S. jello* Cuvier, 1829	1	2%
140	*S. qenie* Klunzinger, 1870	1	2%
XXIII	**LABRIDAE**		
141	*Cheilinus fasciatus* (Bloch, 1791)	44	88%
142	*C. trilobatus* Lacepède, 1802	2	4%
143	*C. undulatus* Rüppell, 1835	14	28%
144	*Choerodon anchorago* (Bloch, 1791)	6	12%
145	*C. fasciatus* (Günther, 1867)	1	2%
146	*Coris aygula* Lacepède, 1802	1	2%
147	*Epibulus insidiator* (Pallas, 1770)	35	70%
148	*Oxycheilinus diagrammus* (Lacepède, 1802)	16	32%
XXIV	**SCARIDAE**		
149	*Bolbometopon muricatum* (Valenciennes, 1840)	2	4%
150	*Calotomus carolinus* (Valenciennes, 1839)	1	2%
151	*Cetoscarus bicolor* (Rüppell, 1828)	33	66%
152	*Chlorurus bleekeri* (de Beaufort, 1940)	45	90%
153	*C. microrhinos* (Bleeker, 1854)	12	24%
154	*C. sordidus* (Forsskål, 1775)	36	72%
155	*Hipposcarus longiceps* (Bleeker, 1862)	20	40%
156	*Scarus altipinnis* (Steindachner, 1879)	8	16%
157	*S. chameleon* Choat and Randall, 1986	7	14%
158	*S. dimidiatus* Bleeker, 1859	36	72%
159	*S. flavipectoralis* Schultz, 1958	43	86%
160	*S. forsteni* (Bleeker, 1861)	5	10%

continued

No.	Family/species	No. sites where present	Percent occurrence
161	*S. frenatus* Lacepède, 1802	1	2%
162	*S. ghobban* Forsskål, 1775	6	12%
163	*S. globiceps* Valenciennes, 1840	2	4%
164	*S. niger* Forsskål, 1775	49	98%
165	*S. oviceps* Valenciennes, 1839	4	8%
166	*S. psittacus* Forsskål, 1775	5	10%
167	*S. pyrrhurus* (Jordan and Seale, 1906)	1	2%
168	*S. quoyi* Valenciennes, 1840	4	8%
169	*S. rivulatus* Valenciennes, 1840	1	2%
170	*S. rubroviolaceus* Bleeker, 1849	6	12%
171	*S. schlegeli* (Bleeker, 1861)	4	8%
172	*S. spinus* (Kner, 1868)	14	28%
XXV	**ACANTHURIDAE**		
173	*Acanthurus blochi* Valenciennes, 1835	7	14%
174	*A. dussumieri* Valenciennes, 1835	8	16%
175	*A. fowleri* de Beaufort, 1951	5	10%
176	*A. lineatus* (Linnaeus, 1758)	3	6%
177	*A. mata* (Cuvier, 1829)	6	12%
178	*A. nigricans* (Linnaeus, 1758)	3	6%
179	*A. nigricaudus* Duncker and Mohr, 1929	26	52%
180	*A. nigrofuscus* (Forsskål, 1775)	1	2%
181	*A. nubilus* (Fowler and Bean, 1929)	2	4%
182	*A. olivaceus* Bloch and Schneider, 1801	3	6%
183	*A. pyroferus* Kittlitz, 1834	39	78%
184	*A. thompsoni* (Fowler, 1923)	11	22%
185	*A. xanthopterus* Valenciennes, 1835	6	12%
186	*Ctenochaetus* spp.	50	100%
187	*Naso annulatus* (Quoy and Gaimard, 1825)	7	14%
188	*N. brachycentron* (Valenciennes, 1835)	2	4%
189	*N. brevirostris* (Valenciennes, 1835)	11	22%
190	*N. hexacanthus* (Bleeker, 1855)	10	20%
191	*N. lituratus* (Bloch and Schneider, 1801)	40	80%
192	*N. lopezi* Herre, 1927	1	2%

No.	Family/species	No. sites where present	Percent occurrence
193	*N. thynnoides* (Valenciennes, 1835)	1	2%
194	*N. tuberosus* Lacepède, 1801	3	6%
195	*N. unicornis* Forsskål, 1775	6	12%
196	*N. vlamingii* Valenciennes, 1835	25	50%
197	*Zebrasoma veliferum* Bloch, 1797	34	68%
XXVI	**SIGANIDAE**		
198	*Siganus argenteus* (Quoy and Gaimard, 1824)	3	6%
199	*S. corallinus* (Valenciennes, 1835)	24	48%
200	*S. doliatus* Cuvier, 1830	18	36%
201	*S. lineatus* (Linnaeus, 1835)	4	8%
202	*S. puellus* (Schlegel, 1852)	29	58%
203	*S. punctatissimus* Fowler and Bean, 1929	25	50%
204	*S. punctatus* (Forster, 1801)	7	14%
205	*S. vulpinus* (Schlegel and Müller, 1844)	32	64%
XXVII	**SCOMBRIDAE**		
206	*Grammatorcynos bilineatus* (Quoy and Gaimard, 1824)	6	12%
207	*Gymnosarda unicolor* (Rüppell, 1836)	6	12%
208	*Rastrelliger kanagurta* (Cuvier, 1816)	1	2%
209	*Scomberomorus commerson* (Lacepède, 1800)	1	2%

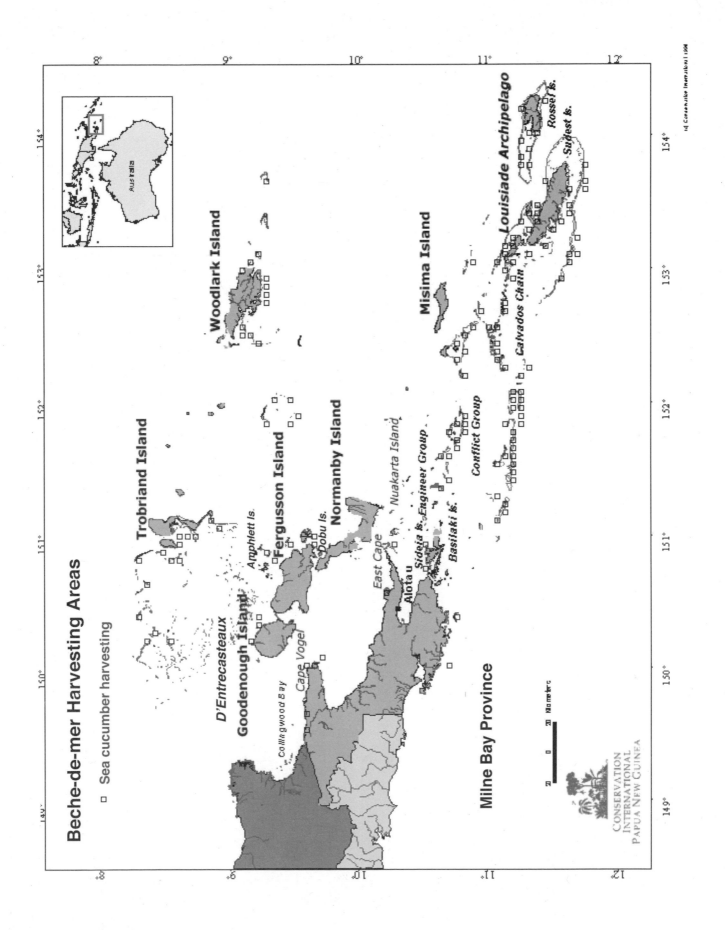

Beche-de-mer Harvesting Areas

□ Sea cucumber harvesting

Trobriand Island

Woodlark Island

D'Entrecasteaux

Goodenough Island

Fergusson Island

Normanby Island

Misima Island

Louisiade Archipelago

Rossel Is.

Sudest Is.

Calvados Chain

Conflict Group

Nuakarta Island

Engineer Group

Sideia Is.

Basilaki Is.

Dobu Is.

Amphlett Is.

East Cape

Alotau

Cape Vogel

Collingwood Bay

Milne Bay Province

Kilometers

Australia

Appendix 7

List of fishes caught by Brooker islanders using various techniques

Jeff P. Kinch

Misima name	Scientific name
TROLLING	
yalyal	*Grammatorcynos bilineatus*
suwa	*Elegatis bipinnulata*
tuna	Scombridae
yellowfin tuna	*Thunnus albacares*
utul	*Aprion virescens*
trevally	Carangidae
barricuda	Sphyraenidae
kibukibu	*Acanthocybium solandri*
mackeral	Scombridae
leplepa	*Megalaspis cordyle*
kingfish	*Seriola* spp. (Carangidae)
maimua	*Grammatorcynos bicarinatus*
sharks	Carcharhinidae
enipola	*Lutjanus bohar*
NETTING	
atuni	*Selar boops*
pilihul	Mullidae
mullet	Mugilidae
vivilal	*Siganus* sp.
boxfish	Ostraciidae
tuna	Scombridae
yalyal	*Grammatorcynos bilineatus*
suwa	*Elegatis bipinnulata*
salasala	*Scarus* spp.
legalegal	*Kyphosus* spp.
batfish	*Platax* spp.
ahiat	*Lethrinus erythracanthus*
yatela	*Naso unicornis*
buhumanwi	*Lethrinus olivaceus*
trevally	Carangidae
dayaya	*Rastrelliger kanagurta*

Misima name	Scientific name
vanavana	*Pentaprion longimanus*
gasawa	*Lethrinus harak*
pepeka	Pomacanthidae/Chaetodontidae
niuniu	*Upeneus vittatus*
gawagawa	*Decapterus russelli*
malawi	*Acanthurus dussumieri*
anuwal	*Sphyraena* spp.
kakawola	*Epinephelus merra*
leu	Belonidae
utul	*Aprion virescens*
magoga	*Alectis ciliaris*
yesimoli	*Naso tuberosus*
mackeral	Scombridae
labeta	*Lethrinus nebulosus*
tokeli	Lethrinus spp.
enipola	*Lutjanus bohar*
nwabelele	*Plectrorhinchus* spp.
maninuya	Serranidae
ulihela	*Variola louti*
lamwavalval	*Acanthurus triostegus*
kokok	*Trachinotus blochii*
togoba	*Bolbometopon muricatum*
tolobil	*Thalassoma hardwickei*
taipwehe	*Naso brachycentron*
ganagana	*Lethrinus* sp.
SPEARING	
atuni	*Selar boops*
boxfish	Ostracidae
pilihul	Mullidae
talian	Atherinidae
kabela	*Acanthurus auranticavus*

continued

Misima name	Scientific name
mullet	Mugilidae
kiton	*Siganus* spp.
malawi	*Acanthurus dussumieri*
taipehepehe	*Naso brachycentron*
yatela	*Naso unicornis*
yesimoli	*Naso tuberosus*
maninuya	Serranidae
cardinal fish	Apogonidae
yabwau	*Lutjanus gibbus*
parrotfish	Scariidae
surgeonfish	Acanthuridae
anuwal	*Sphyraena* spp.
legalegal	*Kyphosus* spp.
vivilal	*Signaus* sp.
havani	*Acanthurus nigricauda*
kakawola	*Epinephelus merra*
koil	*Acanthurus lineatus*
ganagana	*Lethrinus* sp.
kakabela	*Acanthurus leucocheilus*
FISHING	
myahmul	*Monotaxis grandoculus*
yabwau	*Lutjanus gibbus*
anuwal	*Sphyraena* spp.
labeta	*Lethrinus nebulosus*
tupatupa	Carangidae
enipola	*Lutjanus bohar*
maninuya	Serranidae
ganagana	*Lethrinus* sp.
bwania	*Epinephelus fuscoguttatus*
gasawa	*Lethrinus harak*
bwalioga	*Choerodon anchorago*
kosa	*Scarus* spp.
tayaka	Apogonidae
kakawola	*Epinephelus merra*
makimaki	*Chelinus* spp.
atuni	*Selar boops*
ahiat	*Lethrinus erythracanthus*
mwalimwaligan	Platycephalidae
siusiu	*Naso brachycentron*
utul	*Aprion virescens*
tokeli	*Lethrinus* spp.
kimakimaga	*Paracanthurus hepatus*
ulisiai	*Lethrinus miniatus*
ulihela	*Variola louti*
tuna	Scombridae

Misima name	Scientific name
shark	Charcharinidae
kibkib/kisep	Balistidae
wanin	*Hologymnosus doliatus*
tatan	*Lutjanus kasmira/quinquelineatus*
uliyapuyapu	*Diagramma pictum*
buhumanawi	*Lethrinus olivaceus*
stingray	Dasyatidae
suwa	*Elegatis bipinnulatus*
lepalepa	*Megalaspis cordyle*
talian	Atherinidae
bwaligila	*Scolopsis xenochrous*
tiger Shark	*Galeocerdo cuvieri*
longtoms	Belonidae
gibala	*Fistularia commersonii*
Napolean	*Cheilinus undulatus*
tawiya	*Plectropomus* spp.
hopahopa	*Plectrorhinchus chaetodontoides*
veya	*Epinephelus lanceolatus*
itoito	*Naso hexacanthus*
nabwalele	*Plectrorhinchus* spp.
yesimoli	*Naso tuberosus*
yatela	*Naso unicornis*
kabela	*Acanthurus auranticavus*
kiton	*Siganus* spp.
havani	*Acanthurus nigricauda*
vivilal	*Siganus* sp.
mullet	Mugilidae
salasala	*Scarus* spp.
togoleli	*Naso lituratus*
boxfish	Ostracidae

Map 1

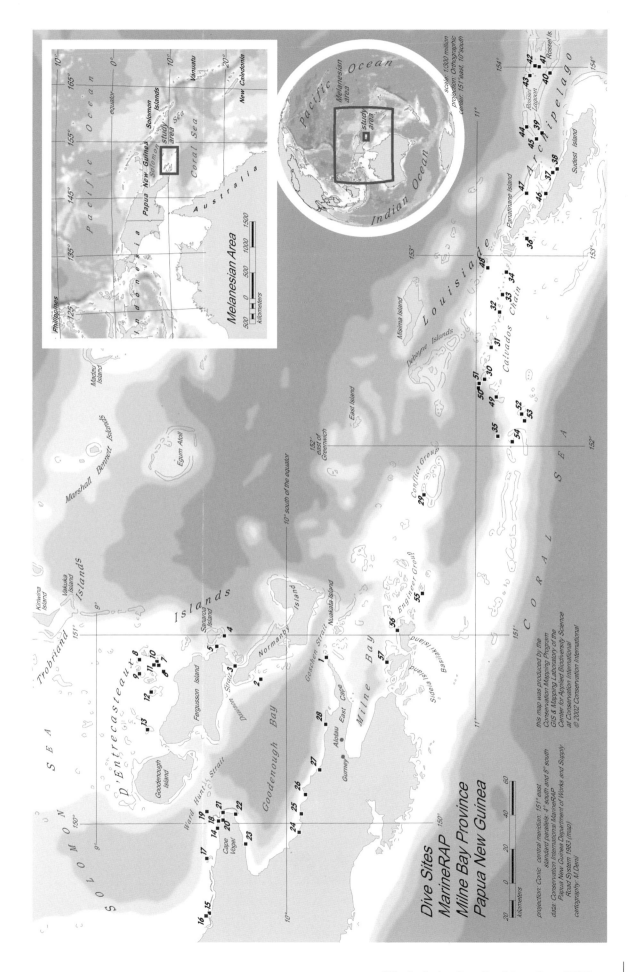

Dive Sites
MarineRAP
Milne Bay Province
Papua New Guinea

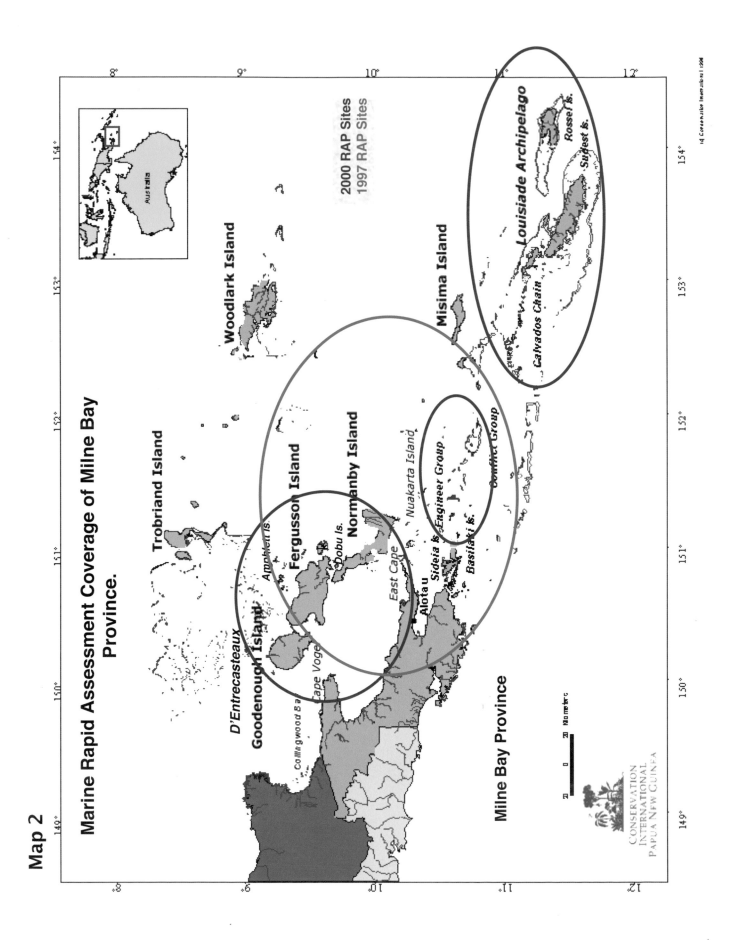

Map 2

Marine Rapid Assessment Coverage of Milne Bay Province.

A raised Miocene reef is now visible as a weathered limestone shelf at Misima Island, typical of some of the islands found in Milne Bay Province.

Commonwealth Scientific and Industrial Research Organisation (CSIRO) fisheries ecologist Darren Dennis with a catch of long-nose emperor (*Lethrinus olivaceus*) on last year's NFA/CSIRO/CI stock assessment and biogeographical survey.

Mortuary prestation or *hagali*. Feasting and mortuary obligations are an important cultural element to the lives of people of Milne Bay Province.

Lake formed from water seepage and run-off in the open-cut pit of Misima Mines Limited, Misima Island. Misima Mines Limited is due for closure in 2004.

Measuring trochus for sale at a local trade store. The minimum size limit for trochus is 8 cm.

Mine road and rehabilitation work at Misima Mines Limited. The islands of the West Calvados Chain are visible in the background.

Boys and their toy sailing canoes. By using toys they learn the dynamics of sailing and wind direction for practical use later as adults when they will either skipper or crew sailing canoes.

A beche-de-mer catch consisting mostly of Amberfish (*Thelenota anax*) and Curryfish (*Stichopus hermanni*, previously *variegatus*). One *Tridacna gigas* is also seen.

View of the main village *Awan bwabwatana* at Brooker Island.

Women cleaning giant clam. The muscle from giant clam species *Tridacna gigas*, *T. derasa*, and *T. maxima* were previously sold for export. The smaller *Hippopus hippopus* is predominantly taken for subsistence and trade.

In-laws' obligation or *muli* for a mortuary presentation.

The Tuna longlining vessel *Ko Shinsei* (PNG 44) run aground on the reefs at Panadaludalu Island in the West Calvados Chain. Illegal entry by longliners within the 6-mile zone is a concern for many villagers in Milne Bay Province.

Sailing canoe returned with sago thatch for a new house under construction. Subsistence trading promotes regional economic specialization based upon ecological constraints, as well as being an important part of the ceremonial and social life of the people of Milne Bay Province.